Island Studies Series

Series Editor
Lino Briguglio

Banking and Finance in Islands and Small States

Island Studies Series

Series Editor
Lino Briguglio
Islands and Small States Institute, Foundation for International Studies, University of Malta

The Island Studies series focuses on issues which particularly affect small inhabited islands, with special reference to politically independent ones.

The study of islands, as distinct from other geographical entities, is developing as a special area of interest because islands tend to face special problems associated with smallness, insularity, fragile eco-systems and proneness to natural disasters, which render them very vulnerable in the face of forces outside their control. This condition sometimes threatens their very economic viability.

Small island states also offer unique perspectives for study, especially in the areas of tourism and leisure, geography, anthropology, sociology, economics, the environment and sustainable development in general.

The volumes in this series include studies of a general nature and case studies, aimed at scholars and practitioners engaged in the study and in the management of small islands.

Also published in the same series:

Sustainable Tourism in Islands and Small States: Issues and Policies Edited by Lino Briguglio, Brian Archer, Jafar Jafari and Geoffrey Wall.

Sustainable Tourism in Islands and Small States: Case Studies Edited by Lino Briguglio, Richard Butler, David Harrison and Walter Leal Filho

Banking and Finance in Islands and Small States

Edited by
Michael Bowe
Lino Briguglio
James W. Dean

PINTER

London and Washington

First published 1998 by
Pinter, A Cassell Imprint
Wellington House, 125 Strand, London WC2R 0BB
PO Box 605, Herndon, VA 20172

British Library Cataloguing in Publication Data

Banking and finance in islands and small states
   1. Finance   2. Banks and banking
   I. Bowe, Michael   II. Briguglio, Lino   III. Dean, James, W.
332

   ISBN 1855674890

Library of Congress Cataloging-in-Publication Data

Banking and finance in islands and small states/edited by
   Michael Bowe, Lino Briguglio, James W. Dean.
       p. cm. – (Island studies series)
   A selection of papers presented at the International Conference on
Banking and Finance in Islands and Small States, organized by the
Islands and Small States Institute of the Foundation for
International Studies at the University of Malta in collaboration
with the Centre for International Business Studies at Pace
University, and held January 1995 in Malta.
   Includes bibliographical references and indexes.
   ISBN 1–85567–489–0 (hardcover)
   1. Banks and banking, International–Congresses. 2. States,
Small–Economic conditions–Congresses. 3. Islands–Congresses.
I. Bowe, Michael. II. Briguglio, Lino. III. Dean, James W.
IV. International Conference on Banking and Finance in Islands and
Small States (1995 : Valetta, Malta). V. Series: Island studies
series (London, England)
HG205.B36 1997
332. 1 – dc21   96–54223
                                                     CIP

Typeset by York House Typographic Ltd, London

Printed and bound in Great Britain by Biddles Limited,
Guildford and King's Lynn

Contents

Contributors

Brian Archer is a Professor in the Department of Management Studies, University of Surrey, England, where for seven years he was a Pro-Vice-Chancellor. Prior to joining Surrey in 1978 he was Director of the Institute of Economic Research in the University College of North Wales, Bangor. Over the past 25 years he has compiled over 50 input–output tables for regions and localities in the United Kingdom and for overseas countries. He specializes in tourism economics.

Paul V. Azzopardi graduated in Accountancy at the University of Malta in 1983. He became a member of the Malta Institute of Accountants in 1983 and a Certified Public Accountant in 1984. He studied investment and finance for two years at the University of British Columbia, Vancouver, Canada, where he obtained a first-class Master of Business Administration (MBA) degree. He worked at the Malta Development Corporation between 1979 and 1988 and acted as director of two companies with government shareholding. He lectured at the University of Malta in accountancy, finance and investments. In 1989 he set up his own financial and investment services firm. The group of companies includes Azzopardi Stockbrokers Ltd, Azzopardi Investment Management Ltd and Azzopardi Insurance Brokers Ltd.

Michael Bowe is Director of the Chartered Institute of Bankers Centre, Manchester School of Management, UMIST. He was formerly (1984–1993) Assistant Professor of Economics, Department of Economics, Simon Fraser University, Canada. His consulting experience is extensive. Clients include the World Bank, the Canadian International Development Agency, the Government of British Columbia, Union Bank of Finland Ltd, Den Danske

Bank, Interbank Ltd, Kansallis-Osake-Pankki, the Singapore Stock Exchange and the Finnish Options Market. He has published extensively on themes relating to banking and finance. His recent research publications include 'Has the market solved the debt crisis?', *Princeton Essays in International Finance*, 1997; 'Debt–equity swaps and the enforcement of sovereign loan contracts' with James Dean, *Journal of International Development*, 1996; 'The costs of arbitrage and futures market trading activity', *International Journal of the Economics of Business*, 1994; and 'Debt–equity swaps: investment incentive effects and secondary market prices' with James Dean, *Oxford Economic Papers*, 1993. He is the author of *Eurobonds* (Dow Jones Irwin, Homewood, Illinois, 1989; 2nd edition forthcoming, spring 1997).

Lino Briguglio is Professor of Economics at the University of Malta, Director of the Islands and Small States Institute of the Foundation for International Studies and Director of the University Centre on the Island of Gozo. He was awarded a PhD in Economics from the University of Exeter (UK) and a Special Diploma in Social Studies from the University of Oxford. He obtained his honours and master's degrees in Economics from the University of Malta. His main area of interest is islands and small states studies. He is the author of three books on the Maltese economy, and was also the editor of a special issue of *World Development* (Oxford: Pergamon Press, February 1993) on 'Islands and Small States' and of two volumes entitled *Sustainable Tourism in Islands and Small States* (London: Pinter, March 1996). He has also edited three volumes on issues related to tourism, history and culture on the island of Gozo. He is the editor of *Bank of Valletta Review*, published by the Bank of Valletta, and co-editor of *Insula – International Journal of Island Affairs*, published in collaboration with UNESCO. He also serves on the editorial board of a number of other journals. Professor Briguglio has published a large number of articles on islands and small states, one of which relates to the economic vulnerabilities of small islands states, in *World Development* (vol. 23, no. 9, 1995). Professor Briguglio has acted as consultant to several international organizations including UNCTAD and CARICOM. He has also represented the Maltese government in United Nations meetings dealing with island states affairs, including the UN Global Conference on the Sustainable Development of Small Island States.

Stephen Carse was born in Liverpool in 1953, living there until entering Cardiff University in Wales in 1971. He graduated with an upper second class honours degree in economics in 1974 and went on to gain a MSc in Economics. From 1975 to 1977 he was a Research Associate at Warwick University, working alongside Professors John Williamson, Geoffrey E. Wood and the late Fred Hirsch in a study of the relevance of exchange controls in the context of international financing procedures in UK commerce. He then moved to take up a lectureship post in economics at Leicester

Polytechnic (now De Monfort University) in England while maintaining research interests in foreign trade and industrial development. Past publications include *Financing Procedures in British Foreign Trade* (Cambridge University Press, 1980) co-authored with Geoffrey Wood and John Williamson. After a brief spell in business consultancy Mr Carse took up the post of Economic Adviser for the Isle of Man government in January 1989. His responsibilities include the provision of economic, socio-economic and financial advice affecting any issue of government policy, and the running of a division of the government Treasury concerned primarily with the collection and analysis of economic statistics.

Marcel Cassar holds an MBA from the University of Wales (in cooperation with Manchester Business School) and is a certified public accountant and auditor, having practised in the profession with Price Waterhouse in Malta and Italy. Since 1991 he has played an active and key role in the development of a business and regulatory framework for Malta as a centre for offshore and international banking activities. In that capacity he has represented Malta in the Offshore Group of Banking Supervisors, which meets under the auspices of the Basle Committee on Banking Supervision. Until the end of 1996, Mr Cassar was Head of Offshore Banking Supervision, Deputy Director of Investment Services and a member of the Executive Committee at the Malta Financial Services Centre. He is also a lecturer in banking studies and banking regulation at the University of Malta. His special area of interest, which was also the research theme for his MBA dissertation, is prudential supervision of state-owned banking firms.

James W. Dean is Professor of Economics at Simon Fraser University in Vancouver, British Columbia, Canada, and Kaiser Professor of International Business at Western Washington University in Bellingham, Washington, USA. From 1993 to 1996, he was Bank of Valetta Visiting Professor of International Banking at the University of Malta. He holds a BSc degree in Mathematics from Carleton University, and MA and PhD degrees in Economics from Harvard University. Professor Dean publishes in scholarly journals, books and the popular press on international finance, domestic and international banking, and macroeconomics. For the past several years, his research has focused on international sovereign debt and debt relief, and international capital flows. He consults on these and related issues for the Canadian International Development Agency, the World Bank and other agencies. Professor Dean has held appointments at many institutions world-wide, including Columbia University, New York University, the University of Toronto, the University of Paris I, the Institut d'Études Politiques (Paris), INSEAD (Fontainebleau), the Helsinki School of Economics, the University of Malta, the University of Cape Town, the University of Hong Kong, the

Institute of Southeast Asian Studies (Singapore) and the University of Manchester Institute of Science and Technology (UMIST).

Joe Falzon holds a PhD in economics from Northwestern University, Evanston, Illinois, where he was a Fellow of Rotary International. He has taught at Roosevelt University, Chicago (1982–84), Northwestern University (1985–86) and the University of Cincinnati (1986–87). Since 1988 he has been teaching at the University of Malta and has also served as a consultant to the government of Malta. He is the author of several papers and research projects on different economic issues and is currently Head of the Department of Banking and Finance at the University of Malta.

Bruce Felmingham is currently Reader in Economics at the University of Tasmania. His teaching and research interests are focused on international finance, monetary and labour economics and federal, state and local government finance. He is the author or co-author of several books, book chapters and research monographs in addition to 50-odd research journal articles, articles in conference proceedings and papers. Dr Felmingham has worked as a consultant/adviser to federal, state and local government agencies in several private firms and trade unions. He is a well-known media commentator on Tasmanian public issues and has participated in many public inquiries. His research and consultancy activities are based on expertise in mathematical modelling, time series analysis and operations research. He is currently Chairman of the Board of TasVest Ltd, a local merchant bank.

Maximilian J.B. Hall graduated with a first-class honours degree in Economics from Nottingham University in 1975. He received a PhD from the same university in 1978. He joined the staff of the Economics Department at Loughborough University in 1977 and is currently a Senior Lecturer in that Department. He has published seven books (one co-authored) in the areas of money, banking and finance. He has also contributed over 60 articles to academic and professional journals, and chapters to a further ten books. In addition, he has three entries in the *New Palgrave Dictionary of Money and Finance* published by Macmillan in 1992. His current research activities embrace UK banking supervision, central and commercial banking developments in the UK, USA, Japan and the EU, and financial regulatory issues in general. Dr Hall has carried out a number of consultancy projects for public- and private-sector organizations and has contributed on the Top Management Programme run by the Cabinet Office. He has also delivered over 100 papers at conferences around the world. Apart from teaching at Loughborough University, Dr Hall has held visiting lectureships at Leicester University, Nottingham University, Sheffield University (Management School), the City University Business School, University of Malta and the Hong Kong and Shanghai Banking Corporation (London office). He has also acted as Economic Adviser to the Macau Business Centre since 1982.

Mark P. Hampton is Senior Lecturer in Modern Economic History in the Department of Economics, University of Portsmouth. Dr Hampton was educated in Jersey and at the University of East Anglia, Norwich. He has been an invited speaker at international conferences in North America, East Asia and Europe. He has written several academic journal articles and edited book chapters. His new book, *The Offshore Interface: Tax Havens in the Global Economy*, is published by Macmillan. At present, Dr Hampton serves on the Executive Committee of the International Small Islands Studies Association. In addition he is a member of the Development Studies Association, the Association of Southeast Asian Studies in the UK (ASEASUK) and the Economic History Society.

T.K. Jayaraman, formerly senior economist in the South Pacific Regional Mission of the Asian Development Bank (ADB), is currently a fellow at the MacMillan Brown Centre for Pacific Studies, University of Canterbury, New Zealand. After obtaining his BA (Hons) degree in Economics from Madras University, India, he joined the Indian Administrative Service in 1960. After working at both policy formulation and implementation levels, he proceeded to earn his master's and PhD degrees from the University of Hawaii in 1975. He is the recipient of an East–West Center Grant (1968–69) and Fulbright Grant (1972–75). He was also a Doctoral Intern at the East–West Center during 1975. He taught economics at the University of Hawaii and was part-time lecturer at Gujarat University during 1976–82. He joined ADB in 1982 and since then he has processed several loan and technical assistance projects in the areas of agriculture and rural development in Fiji, the Solomon Islands, Tonga and Western Samoa, all in the South Pacific. Since 1992, he has been involved in economic and sector work and has prepared policy-oriented papers relating to South Pacific island economies. He has published extensively on the South Pacific islands. Presently, in addition to his work at the Bank, he is a part-time lecturer in economics at the University of the South Pacific's Extension Centre at Port Vila, Vanuatu.

Brian Kettell has worked for Citibank, American Express and Shearson Lehman, and was Director of Training and Assistant Vice President at the Arab Banking Corporation in Bahrain. He has published several books on international financial markets, including *The Finance of International Business, A Business Man's Guide to the Foreign Exchange Market, The International Debt Game, The Foreign Exchange Handbook* and *Monetary Economics* and *Gold*. He has taught at several universities and published over 70 articles in journals, business magazines and the financial press. He is currently a freelance financial markets trainer. He is editing and writing a new series of books on financial markets to be published in 1998 by the *Financial Times*.

Mervyn K. Lewis is National Australia Bank Professor in the School of Economics, Finance and Property at the University of South Australia.

University of Nottingham. He has been a consultant to the Australian Financial System Inquiry, a Visiting Scholar at the Bank of England, and a visiting professor at the University of Cambridge and the Wirtschaftsuniversität Wien. Professor Lewis has published ten books and numerous articles on monetary economics, contributing eleven entries to *The New Palgrave Dictionary of Money and Finance*. He is on the editorial boards of *Accounting Forum, Bank of Valetta Review, The Review of Policy Issues* and *Accounting and Auditing: Islamic Perspective*. In 1986 he was elected a Fellow of the Academy of the Social Sciences in Australia, one of the four learned academies in Australia. His most recent book is *The Australian Financial System. Evolution, Policy and Practice* (Addison Wesley Longman, 1997).

Sir Indur Ramphul studied Public Administration at the University of Exeter in the UK from 1962 to 1964. He attended Central Banking courses at the Bank of England in 1969 and the Federal Reserve Bank of New York in 1976. Sir Indur is a leading and highly respected figure in Mauritian banking and financial circles. After a career of 16 years at the Accountant General Department and at the Ministry of Finance, he joined the Bank of Mauritius in 1967 as a Manager. He was successively promoted to Chief Manager, Managing Director and finally, in June 1982, to the position of Governor of the Bank of Mauritius, which he occupied until 1996. He also acted as alternate Governor for Mauritius of the International Monetary Fund. Sir Indur has played a key role in the development of the banking sector in Mauritius and, more recently, in the modernization of the country's banking legislation. He is also the driving force behind offshore banking development in Mauritius. Until 1996 he was the Chairman of the Offshore Banking Technical Committee set up by the government to look into all aspects of offshore banking in Mauritius including the grant of offshore banking licences. It was in recognition of his invaluable services to banking in Mauritius that he was knighted on 31 December 1990 by Her Majesty Queen Elizabeth II on the recommendation of the Prime Minister of Mauritius.

Eugene Sarver is an Associate at Intercap Investments Inc. (New York), where he is Director of its *Monitor of Islamic Funds* service and also conducts global marketing of derivative financial products. Previously he was an Associate Professor at the Lubin School of Business-Graduate Division of Pace University, New York City and an economist for Chase Manhattan and Crédit Lyonnais. Concurrently, he serves on the Board of Directors of the American Heritage Fund and the American Heritage Growth Fund. He is chairman of the quarterly (NYC, Chicago, San Francisco) 'Fundamentals of Foreign Exchange' and the semi-annual 'Introduction to International Financial Markets' programmes of the World Trade Institute of the Port Authority of New York and New Jersey, and serves as consultant to global financial firms including Daiwa, the Republic National Bank of New York and

Nomura, and governmental entities including MIDA (Malaysia) and INFOR-MASI (Indonesia). He arbitrates commercial bank disputes for the American Arbitration Association and regularly conducts financial training programmes for the US State Department through its Barents and PIET affiliate for premier financial institutions including Citibank, Swiss Bank Corporation, Chase Manhattan and Merrill Lynch, and for Russian-American companies outside the United States. Dr Sarver is the author of *The Eurocurrency Market Handbook* (New York: Simon & Schuster/Prentice Hall, 2nd edn, 1992) and has written numerous articles on international finance and economics based on his research in 90 countries. Currently he writes the monthly foreign exchange report, including analyses and forecasts, for *World Trade* magazine.

Paul Styger is Associate Professor in the Department of Economics, Money and Banking at the Potchefstroom University for Christian Higher Education, Potchefstroom, South Africa. He received his master's degree from the University of the Free State, and his doctorate in Economics from the Potchefstroom University for CHE. He has done research for and serves as a consultant to banks and government departments. Professor Styger has contributed chapters to five books and was editor or co-editor of six books. He has published various academic and popular articles. He is a member of the National Council of the Economics Society of South Africa as well as the Development Society of Southern Africa.

Anthony B. van Fossen is a Senior Lecturer in the School of Australian and Comparative Studies at Griffith University in Brisbane, Australia. He is primarily interested in the international political economy of the Pacific islands and is engaged in a full-scale study of the region's offshore financial centres. His publications include *The International Political Economy of Pacific Islands Flags of Convenience* (1992) and, more recently, 'Corporate power in the Pacific islands' (*Current Sociology*, 1995).

Theo van Wyk graduated from Potchefstroom University for Christian Higher Education, South Africa. He was economist and asset and liability manager at Saambou Building Society. He then joined SPL World Group (SA) as Marketing Manager, Banking and Corporate. Presently he is senior consultant Group ALCO at ABSA bank.

Preface

Many small islands, some of which are independent states, have a relatively large financial sector and serve as offshore financial centres in different parts of the world. The Caribbean and the Asia-Pacific regions are dotted with islands that have had considerable success in this regard. Several islands and small states in Europe have also made a name for themselves in financial services. This phenomenon prompted the Islands and Small States Institute of the Foundation for International Studies at University of Malta to organize the International Conference on Banking and Finance in Islands and Small States, in collaboration with the Centre for International Business Studies at Pace University. The conference, which was held in Malta in January 1995, attracted scholars and practitioners from all parts of the world. The chapters in this volume represent a selection of the papers presented at that conference.

Small islands tend to have limited economic possibilities. Reliance on economic activities such as manufacturing and agriculture is often not a viable option, mostly because of size limitations and a lack of natural resources. Many islands have therefore had to seek alternative routes for economic development, with tourism and financial services featuring prominently in this regard.

One can easily understand why many islands succeed in tourism; warm climates and sea, sand and sun endowments attract tourists, and the small scale and insularity factors are not usually major disadvantages in this respect. On the contrary, they typically serve as added attractions.

It is not, however, obvious why many islands have been so successful in developing a financial services sector. This type of economic activity demands very specialized personnel, often a scarce commodity in small islands, and requires that the host country keep pace with global techno-logical developments in financial services. This intriguing issue is discussed in some of the chapters of this book.

An important theme associated with banking and finance relates to scope and incidence of international regulation. As a consequence of recommenda-tions by the Bank for International Settlements, a global change is taking place in regulatory regimes. There are, however, special regulatory con-straints faced by islands and small states, arising from their small size and insularity. This aspect of banking and finance is explored in the book.

Producing an edited book of this type is a challenging task, especially because we, the editors, live in different continents, and the authors hail from all four corners of the world. Our task was facilitated by the support we received from various people. We are very grateful for the contribution made by all the participants of the January 1995 conference, who in submitting their papers enabled us to draw from them the selection we chose for this book. We would like to thank all authors of the papers selected for their prompt replies when they were requested to revise parts of their text. We are also very grateful to Professor Salvino Busuttil, Director-General of the Founda-tion for International Studies, for his help and encouragement.

Finally we would like to acknowledge the excellent support provided by Maryrose Vella and Rose Anne Agius of the Islands and Small States Institute, who handled the large amount of international correspondence that this publication entailed.

Michael Bowe, Lino Briguglio and James W. Dean

Introduction

Michael Bowe, Lino Briguglio and James W. Dean

This volume contains a selection of papers presented at the International Conference on Banking and Finance in Islands and Small States held in Malta in January 1995. The main areas covered in the book relate to the emergence of offshore financial centres (OFCs), in particular the locational and jurisdictional factors that played a part in the development of such centres, the regulatory and deregulatory regimes in small state jurisdictions, and the economic benefits and costs that can be derived from offshore economic activity.

Although none of the chapters of this book attempts to analyse what constitutes small size, and how size can be measured, all of them assume that small economic size can pose serious economic constraints, principally lack of natural resources and inability to exploit economies of scale. This reduces the economic development options for the islands and small states concerned. As Carse argues in Chapter 9, such constraints may have been key factors in forcing many small jurisdictions to make their living in the world at large by 'engineering' a comparative advantage: that is, by using what resources and jurisdiction they have to supply products, such as financial services, for which there is an offshore demand.

Emergence and Evolution of Offshore Centres

The chapters by Kettle, Sarver, Ramphul and Jayaraman comprise a series of case studies with a common theme, namely exploration of the primary factors which lay behind emergence of the six offshore banking centres in Bahrain, Belize, Labuan, Malta, Mauritius and Vanuatu. The studies also provide some tentative predictions as to the likely future evolution and importance of offshore banking activity in the context of the economic development of these islands and small states.

At the macro-economic level, the studies reach a consensus that the most important factors contributing to the successful introduction of offshore financial activity are the rate of economic growth in the region in which the particular centre is located, and the willingness of the centre to become integrated with neighbouring economies in its region. It seems that Labuan, in particular, has received extensive benefits from being situated in the high-growth South-East Asian area, while Mauritius has reaped dividends from its membership of COMESA.

From the micro-economic perspective, specific individual characteristics of the centres have also fostered their development and growth. The most important of these relate to the perceived stability of their political and economic institutions; the presence of special regulatory and fiscal incentives such as the absence of withholding tax on offshore banks and their customers; and the introduction of double taxation treaties. These incentives also include a willingness by local governments to liberalize their financial sectors, most importantly by phasing out their foreign exchange controls; and an ability to provide the infrastructure and human capital necessary to service the offshore sector adequately.

It also became apparent at an early stage in the development of the centres, all of which arrived relatively late on the offshore scene, that their viability and growth prospects substantially derive from their ability to diversify into a wide variety of financial services. This enables them to exploit the benefits of scope economies, particularly in the context of making appropriate use of the human capital skills of financial, legal and technical specialists within the centre.

Thus with the exception of Bahrain, which is primarily an inter-bank operation, and a regional and international wholesale money centre offering some insurance services, the other centres generally offer related offshore services such as trusteeship of offshore trusts, reinsurance, international company registration, ship registration, and offshore asset/fund management. However, as Sarver points out, the centres differ significantly in the schedules they have chosen for initiating the individual elements of their chosen package of offshore services.

Some authors, notably Jayaraman and Carse, explain the emergence of OFCs as one route that small islands had to take to promote economic growth, given that they face serious constraints in the development of other economic activity, such as industry and agriculture. Jayaraman refers to Vanuatu's declining prospects in the world markets for its traditional exports of copra and other primary products. This small state chose to rely upon offshore financial activities along with tourism. Jayaraman argues that empirical examination of the determinants of offshore financial activities reveals that local taxes do not have any significant influence on OFC activities, and that macro-economic and political stability are the chief factors behind the flourishing activities of OFC institutions.

Hampton attempts to develop a theoretical framework to explain the emergence of OFCs in small islands, with special reference to Jersey. He considers offshore finance in relation to wider theories of international banking and the behaviour of transnational corporations. He categorizes two major approaches or world views of offshore finance: that of orthodox economics, and the alternative so-called 'Marxist' approach. The first is based on the extent to which government intervenes, and generally leads to the conclusion that offshore finance can be ascribed to an attempt to circumvent onshore regulation and taxation. Hampton does not consider this approach a good explanation of the origin of OFC development. The alternative approach considers the development of offshore finance as the outcome of a conflict between industry and finance in the onshore economy. Hampton thus sees the key factors determining OFC emergence as rooted in the different roles and degrees of power of two factions of capital, namely financial and industrial. Hampton considers financial capital to be dominant over industrial capital. This novel approach at explaining the emergence of OFCs, based on political economy, suggests that the creation of offshore enclaves with minimal regulation, such as Jersey, is the concrete result of the relative power of financial capital in the mainland.

Effective Regulation

Kettle, Sarver and Ramphul are all keen to point out that the financial sector liberalization to which they allude should not be confused with lax regulatory standards or supervisory controls. In the aftermath of the 1991 BCCI debacle, discussed in the Bowe and Hall chapter, effective consolidated supervision and monitoring of offshore banking institutions became of paramount importance. Bowe and Hall explore regulatory conditions, and compare the European Commission's Capital Adequacy Directive (CAD) with the revised 1995 Basle Committee (BC) proposals as approaches to regulating market risk in the European market for financial services. Their results suggest that given adequate consolidated supervision, regulatory efficiency may be enhanced by supplementing a CAD-type standardized risk measurement framework, on the one hand, with BC-type proprietary, in-house risk management procedures accompanied by market surveillance, on the other hand. A major exception is where the level of an institution's assets, or an inability to implement consolidated supervision, results in a failure to generate the correct incentives for an institution to implement proprietary in-house risk management practices effectively. In this case, imposition of a CAD-type standardized framework may indeed be the appropriate regulatory response.

However, as stated in the chapter by Cassar, certain elements of the Bank for International Settlements (BIS) framework, such as the so-called 'OECD Club rule', have met with mixed reactions, with bankers in islands and small states claiming that the effect of including such provisions in the BIS standards is to introduce market distortions, thereby mitigating the accord's level-playing-field objectives and undermining its usefulness. Given that, as Sarver notes, the competitiveness of the environment in which OFCs operate is a critical determinant of a particular centre's future development and growth prospects, the incentive may exist for institutions to circumvent any regulatory features which are perceived to be distorting competition. This could have a detrimental impact on the operation of the entire OFC network.

The authors all note that a credible commitment to effective banking supervision in accordance with the international risk-based capital standards proposed by the BIS (under the BIS accord) is a necessary condition for inspiring international investor and bank customer confidence. This is in spite of the fact that the accord's provisions, as noted in the papers by Bowe and Hall, and Styger and van Wyk, do not always achieve the regulatory objectives for which they were designed. Several of the islands and small states considered in these case studies have signalled their commitment to the accord by becoming members of two supervisory groups affiliated with the BIS, namely the Arab Committee on Banking Supervision and the Offshore Group of Banking Supervisors.

Deregulation

In many small states, the issue is not how to regulate financial services, but how to deregulate the system. In recent years a wave of deregulation has occurred, affecting financial sectors throughout the world, including those previously associated with government dirigism. For example, up to 1987, direct state controls were rampant in Malta. After that year, a process of liberalization was introduced. Azzopardi and Briguglio chronicle the process of interest rate liberalization in that island, where until 1992, interest rates on lending and borrowing financial institutions were established by the Minister of Finance and rigidly controlled by the government. Interest rates were kept relatively low, and in some years the real interest on one-year fixed deposits was actually negative. Interest rate control was just one of a range of economic restrictions affecting prices, profit margins, foreign exchange transactions and foreign trade. Foreign exchange control in Malta was particularly severe, with offenders threatened with heavy fines and even imprisonment. This notwithstanding, the prevailing low-interest-rate regime gave to rise to very large leakages in pursuit of a higher rate of returns, as well as tax evasion.

Azzopardi and Briguglio discuss the impact of interest rate liberalization and refer to the pros and cons of direct controls. The authors argue in favour of a gradual approach so as to establish the necessary preconditions for an orderly transition. Thus, changes towards a fully liberalized money market in Malta should not be undertaken without due consideration of market imperfections and prevailing supervisory realities. They conclude that liberalization of external capital flows and establishment of a broad market for the Maltese lira remain the main challenges to be faced in coming years.

The issue of deregulation is also discussed by Felmingham and Dean. The objective of their paper is to derive some lessons from the Asia-Pacific financial deregulation experiences and the associated growth of OFCs. The authors review economic trends in the Asia-Pacific region, discussing the growth of the 'Asian tigers' – Singapore, Hong Kong, Malaysia, Taiwan, Thailand and Korea – as well as recent trends in Japan, China, Australia and New Zealand. They find that financial deregulation is furthest advanced in Singapore, Hong Kong and New Zealand, and that the reform process is accelerating in all other countries.

The general view that emerges from Felmingham and Dean's chapter is that the faster-growing economies are the more deregulated ones. Moreoever, Japan's sluggish reform process and its high domestic costs have encouraged the development of Singapore and Hong Kong as OFCs. This would seem to suggest that financial deregulation is a necessary condition for OFC status.

The authors conclude by arguing that islands and small states wishing to learn from the experiences of Australasian countries should take note not only of the benefits of OFC status, notably increased income and employment, but also the costs, which could include lost government revenue and a distorted pattern of income distribution.

Economic Benefits

The benefits which can be derived from offshore economic activity are highlighted by Archer with reference to Bermuda. He shows that offshore companies can be a very important source of income for small islands. Although he makes no attempt to generalize, and his results apply to the islands of Bermuda, his study clearly shows that such companies have an impact on the balance of payments, public-sector revenue and personal income. In addition, such companies also generate considerable employment.

Archer calculates that in recent years this sector created 5,000 jobs directly and 5,500 indirectly. Given that Bermuda has a population of just over 60,000, the relative size of employment creation is quite impressive. Archer uses an input–output model to quantify the effects of international companies. His method can be applied to other islands, but, as he states, the results may not be replicable in other islands, since the outcome depends very much on the extent to which the host country can supply these companies with the goods and services they require.

Location and Jurisdictional Niches

Locational aspects specific to the development of financial services are treated by several authors. As already mentioned, the advantage enjoyed by Labuan as a result of its geographical location is dealt with in Sarver's chapter.

Lewis identifies six core components which make up the provision of financial services. These are financial packaging, delivery systems, production activities, regulation, management information systems and market interface. In the traditional pattern of finance, the six elements were combined and were location-specific. In the modern pattern of finance shaped by market and technological developments over recent decades, these components can be, and are, separated. Lewis identifies a number of factors relevant for locational decisions and competition among financial centres in Europe, distinguishing between traditional, entrepôt, offshore, market and ancillary centres.

Carse explores how a small island, the Isle of Man, has exploited its locational and political conditions to create a 'jurisdictional niche'. The Isle of Man's strategy has been to deploy its constitutional position – in particular its autonomy over domestic financial and taxation matters – alongside its political strengths – notably its stability and its strong government financial balances – to develop a highly successful international financial centre within a highly diversified economy.

The question of financial jurisdiction is also treated in van Fossen's chapter. He argues that while some of the most prominent OFCs in the world are colonies or territories lacking in independence, the success of Pacific Island tax havens, by contrast, generally increases with their degree of sovereignty. Greater sovereignty has allowed them to develop higher levels of financial secrecy and innovation, and effective strategies to escape regulation and taxation by metropolitan countries. British Commonwealth ties have assisted the most dynamic and sovereign island tax havens. Attempts to promote full-service OFCs have been most dramatically thwarted in those Pacific Islands countries which have been controlled by the United States, although Australia and France have also discouraged their development.

Van Fossen shows that the most successful island tax havens have been internally self-regulating and not subservient to a metropolitan power hostile towards such development. He argues, however, that full sovereignty (in the sense of complete political independence) has not necessarily been a crucial advantage, possibly because such autonomy has sometimes attracted shady operators in finance.

The Money Supply Process

An area that warrants research relates to the money supply process in small economies. The relatively large government and foreign sectors in these economies give a special character to the monetary growth process.

Falzon compares the money supply process in two small island states located in the Mediterranean Sea, namely Cyprus and Malta. Both islands assign major importance to their financial sectors. Falzon employs two approaches. First, he concentrates on the liabilities side of the banking sector's balance sheet. He develops a model of the money multiplier – the ratio of broad money (M2) to the monetary base – and looks at the factors which are causing it to change. In Malta, because people began to hold relatively less currency and more bank deposits, the money multiplier doubled between 1983 and 1992. In Cyprus, by contrast, the money multiplier remained relatively unchanged, and for most of the period it was twice as large as that in Malta.

Falzon then considers the asset side of the banking sector's balance sheet, evaluating the money supply process via growth in domestic credit and in net foreign assets. He finds that before 1982, increases in M2 were caused primarily by increases in net foreign assets, whereas after 1986, they were almost exclusively caused by the expansion of domestic credit, influenced to a significant extent by government borrowing. No such structural change occurred in Cyprus. This conclusion would seem to suggest that the money supply process can vary drastically within a single small state over a relatively short period of time (as was the case in Malta) and that monetary growth patterns in small states, even neighbouring ones, can be very different.

Small Size and Vulnerability

As already stated, although this volume focuses on islands and small states, and therefore small economic size is a major theme running through the various chapters, there is no attempt to propose an indicator of small size. There is, however, an underlying presumption that the path chosen by many small jurisdictions is fraught with danger, notably that associated with vulnerability to external forces. An interesting remark put forward by Jayaraman is that, like tourism, which is a relatively large industry in many islands and small states, offshore financial activity is very fragile, owing to its susceptibility to the reputation and integrity of the host jurisdiction.

In addition, the financial sector in islands and small states is exposed to external forces which can very easily destroy it. As Carse puts it,

> offshore centres must ensure compatibility (at least) with larger nations in respect of licensing, regulation and supervision. Hostility towards offshores can also stem from ignorance of their role in facilitating capital movement and

investment, and from notions of 'unfair competition' *vis-à-vis* nations unable to offer low tax rates.

For this reason, excessive dependence by islands and small states on offshore financial centres and projection of small jurisdictions simply as tax havens might not be viable and sustainable avenues to economic development. It might be wiser for islands and small states to attempt to develop well-balanced and diversified economies, and thereby avoid having too many eggs in one fragile basket.

1

Financial Services Location and Competition among Financial Centres in Europe

Mervyn K. Lewis

Competition among financial centres in the European time zone is often depicted, misleadingly, as a battle between London, Paris and, most recently, Frankfurt. This is misleading because it ignores the role of other locations and the nature of financial services operations. There are important national and regional centres in many countries, and also major specialist centres, such as Luxembourg, Edinburgh and Dublin along with the islands and small states such as the Channel Islands, Isle of Man, Liechtenstein, Malta and Monaco. These centres focus on particular activities such as offshore fund management, tax and financial planning, and the provision of a range of ancillary services complementary to other operations including group treasury, insurance and collective investments for the firms in the principal centres. As such, these centres are in direct competition not just with each other but with the international, national and regional locations.

In examining the nature of this competition, this chapter presents a schematic account of the financial firm, in which six core activities – financial packaging, delivery systems, production activities, regulation, management information systems and market interface – are seen to categorize the provision of financial services. These various components used to be carried out in the one location, but market and technological advances in recent decades have made it possible for them to be geographically separated. It is this ability to shift particular services and parts of the production process to different locations which makes for the complexity of financial services location decisions.

The Location of Finance

Location decisions in finance are made by financial enterprises engaged in the provision of financial services, and these location decisions govern the business in financial centres. Our starting point has to be the structure of the financial firm, and Figure 1.1 depicts some of the processes involved in the provision of financial services, in terms of six elements.

1. Financial instruments and claims provided to customers can be regarded as packaging together a number of financial characteristics: currency, term to maturity, interest rate structure, embedded options, counterparty risk, tax obligations.
2. Delivery or front-office activities involve contact with the customer and often require intensive use of human resources when financial products are 'sold' rather than 'bought'. Included in this category are stockbroking and bond transactions, credit evaluation, syndication, investment management.
3. Back-office or production activities are processes of a repetitive nature such as transactions processing, loan application processing and servicing, credit accounting, custodial services.
4. Regulatory environment denotes that this production takes place within an economic setting under a set of regulatory rules and conventions, information provision, taxation arrangements, and political and currency stability.
5. Management information systems embrace human resource management, inventory control, strategic planning, cost control and profitability analysis, risk exposure and management, treasury operations.
6. Market interface involves access to news and market information services, and use of accounting and legal services, while provision of payment

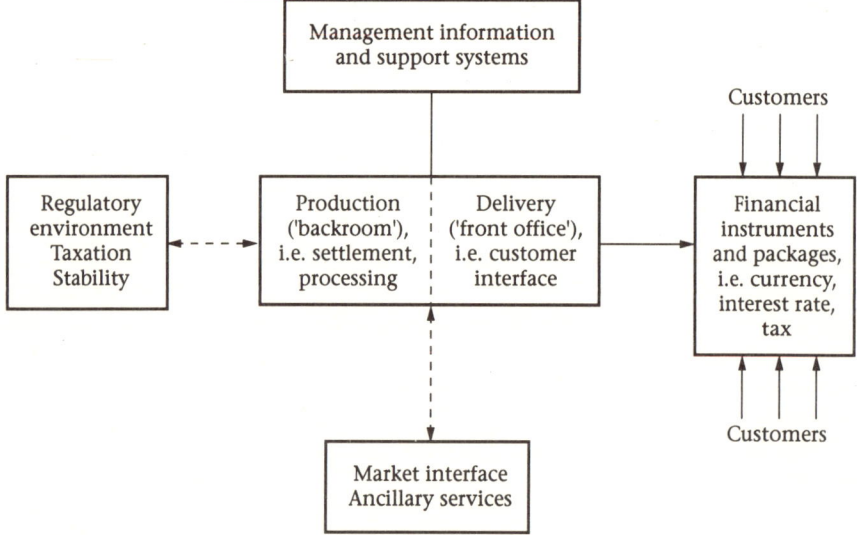

Figure 1.1 Representation of financial services operations

services by banks relies upon correspondent arrangements and inter-bank cooperation through the clearing house.

Types of centre

In the traditional pattern of financing, these elements were combined and were location-specific. Traditional centres, as emphasized by Kindleberger (1974), rose on the back of international trade (London, New York, Chicago, Frankfurt, Hong Kong) and/or balance of payments surpluses (Japan, Switzerland and Germany). The modern pattern of finance has cut these links. It has done so by separating the various elements geographically. As a result, production can be geographically separated from delivery, arranging from legal booking, market interface from portfolio management, support systems from customer servicing. In the process the geography of finance is reshaped. This geographical separation and conversion of location-specific services into 'long-distance' services is not dissimilar to the 'splintering' or 'disembodiment' of services analysed by Bhaghwati (1984) in another context. Technology has played a part in all of this, but so have taxation, regulations and other costs governing the provision of financial services.

Dufey and Giddy (1978) distinguished three types of centre: *capital-exporting centres*, the main role of which is to transfer overseas surplus domestic savings in the traditional way; *entrepôt centres*, which intermediate between international borrowers and lenders and also between international and domestic entities; and *offshore centres*, where intermediation is specifically for non-residents. Based on 'splintering' since then, a further distinction is drawn here in terms of *market centres*, where there is a concentration of trading in particular markets instruments, and *ancillary centres*, which attract various ancillary and support services from the major centres.

Traditional versus Euro centres

Until the 1960s, banks conducted international banking from their home bases. With only minor exceptions, the location of banking and the currency of denomination were inseparable: banking was a natural monopoly of the banks located in the country issuing the currency used for borrowing and lending. Deposits were taken from foreigners, and lending to foreigners occurred, in the currency of the country in which the main offices of banks were located. Much international banking is still carried out in this traditional way. However, only 16 per cent of international bank lending is 'booked out' in this traditional fashion (see Table 1.4 on page 22). Most international lending is instead *externalized* in the Eurocurrency centres, severing the nexus between the location of financial centres and the currency of lending. Before, the major centre (London, New York) was determined primarily by which country had the largest amount of capital for export. London was able to survive the 'dollar hegemony' because of the innovation of the Eurodollar market. In much the same way, London survived the growth of Japanese and German capital exports during the 1980s because of the rapid expansion of the Eurobond market and because of new instruments which blurred the traditional distinctions between international banking and bond markets.

By separating the location of banking from the currency used for borrowing and lending, the Eurocurrency system also separated currency (exchange rate) risk from political risk. In this respect, it altered the financial package in a fundamental way. A depositor holding dollar balances in a country other than the United States combines the currency risk of holding US dollars with the political risk of the particular country (the United Kingdom, Belgium, Singapore, etc.) which is host to the dollar operations. This 'essential feature' was instrumental in the origins of Eurocurrency operations. But growth during the 1960s and 1970s owed more to economic than political factors, for banking had also been separated from the economic risk of taxes and duties levied by the country issuing the currency of denomination. Taxation and economic controls widened the demand for financial services in less regulated locales, while on the supply side banks and other financial intermediaries found that wholesale financial services could be produced externally and delivered 'on the wire'.

This trend to global deposit sourcing and lending in the Eurocurrency markets saw the rise of London and other European cities as international financial centres despite their having relatively weak domestic currencies and lacking the balance of payments surpluses for generating capital exports. In effect, they were free riders on the savings originating in other countries and the monetary stability and financial regulations of those central banks responsible for the provision of international reserve currencies. Success as a financial centre for Eurocurrency banking operations in conjunction with traditional foreign banking revolved around the regulatory environment: the absence of interest rate and exchange controls on foreign currency business, ease of entry and registration procedures, and low reporting requirements.

Offshore locations

Taxation has also influenced the geography of finance, and the favourable tax and business regimes offered by various islands and small states have produced a worldwide proliferation of financial centres (Johns, 1992). The small states in Europe are Andorra, Cyprus, Iceland, Liechtenstein, Luxembourg, Malta, Monaco and San Marino (see *The Banker*, 1992, for the Mediterranean locations). British dependencies in Europe and the Caribbean are also prominent. With the rise of global financing came an associated demand for centres in low-tax, stable economies where fund raisings can be deposited and intra-firm financings made, facilitating cash management and tax planning at an international level. These centres are also bases for investment trusts, captive insurance and international pensions schemes. Many of the operations are of a wholesale nature, closely linked to head office. In some cases the offices established in these centres are merely booking entities, allowing business conducted elsewhere to be legally routed through the offshore location for tax and other benefits. For example, of the 500-odd banks in the Cayman Islands, only about 70 have a physical presence. In this case, what is involved in an offshore operation is a locational separation of loan arranging and other intermediation from booking activities, with the division between the two occurring to a greater or lesser extent according to the nature of the activity and the centre concerned.

Success as an offshore location begins with the domicile having political, economic and fiscal stability and a stable local currency. In addition, it must have solicitors, accountants, stockbrokers and other professionals providing management and financial services, along with good communications in the same time zone as a major centre to allow a link-up with onshore operations. An appropriate mix of financial secrecy, company laws, taxation, accounting requirements, reporting requirements, entry requirements, licensing and regulations is required. The aim is to achieve a long-run reputation for business stability and security of investment. Solvency and prudential oversight are needed for depositor and investor protection. Regulation must be flexible but not too lax, as under-regulation encourages practices which give the centre adverse publicity and taint those using it. In short, the commercial environment must be responsive to the requirements of genuine business, while deterring those seeking to abuse accepted international standards.

Market centres

Banking has always relied on a range of related markets for foreign exchange, for inter-bank funding and for trading bills, bonds and other marketable securities. This requirement has seen banking gravitate to those locales such as London with the 'deepest' markets, able to provide the best liquidity and the finest of margins. These centripetal forces have been reinforced by the growth of international securities activities relative to bank lending (see Figure 1.2), and the associated expansion of derivative markets. In some cases, e.g. Singapore, market growth has built on strength in banking-related business. Chicago and Philadelphia are examples of 'pure' market centres, each with a strong focus upon futures and financial derivatives relative to other aspects of international finance.

Competition among centres has taken the form of the modernization of markets, the application of information technology, and the development of new markets. The modernization of markets in Europe began with Big Bang in London in 1986, closely followed by reforms in Paris, Madrid, Brussels and later Germany as the ability of firms to list on several exchanges saw business migrate across borders. Application of information technology has facilitated

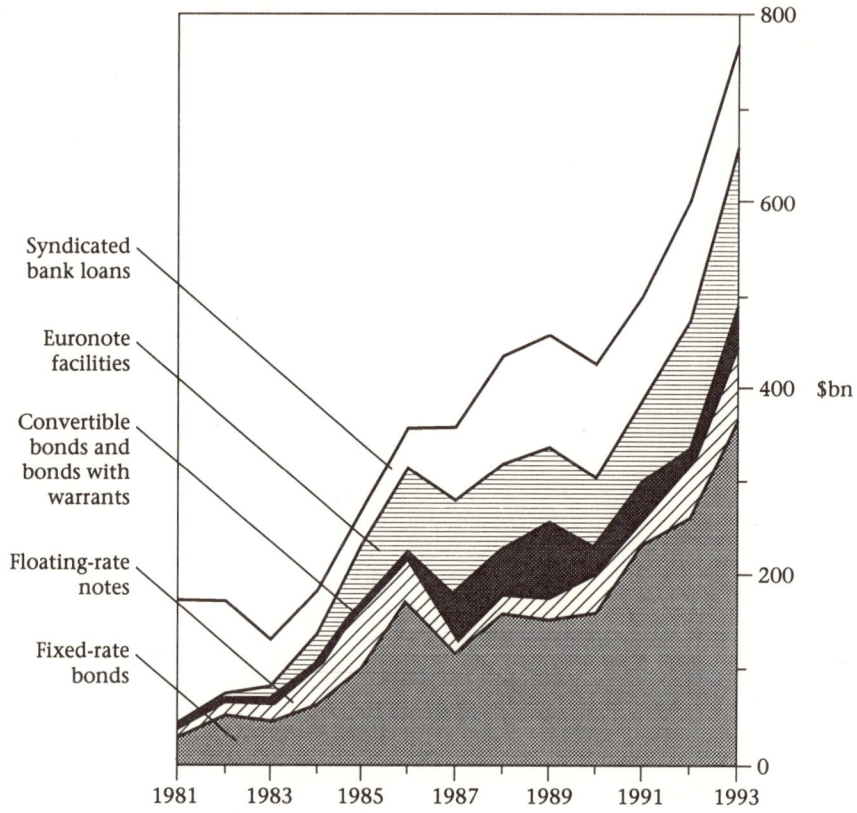

Figure 1.2   International financing activity, 1981–1993
Source: OECD Financial Markets (various)

the cross-border migration of securities business by divorcing portfolio management decisions from the place where the trading is executed, allowing market participants to choose where to transact. This choice of where to undertake trading often involves a decision as to mode of trading system; for example, London's marketmakers *vis-à-vis* the computerized order-matching in Paris. Finally, market innovation has been rapid in terms of derivative markets, with fierce competition between the London International Financial Futures Exchange, the Marché à Terme International de France and the Deutsche Terminbörse. Two features of derivative markets contribute to this competition. Trading in derivatives is often a multiple of the turnover of the underlying cash instruments. Also, the contracts traded are created by the markets and exchanges, which is not the case with cash instruments. Seemingly small variations in contract conditions can have a large impact on derivatives trading. For these reasons, competition in derivatives is often keener than that in spot markets.

Ancillary centres

New technology has altered the way in which the activities shown in Figure 1.1 are carried out, and has facilitated a greater separation of production from delivery in terms of geographic relocation, both within the organization and across firms. Some (e.g. Leveson, 1982) see in this a revolution in integrative services made possible by developments in information technology. Integrative services 'inter-connect firms, units of firms or industries in different stages of the production process or in different locations'. Any service process or product that can be reduced to electronically coded bits of information can be delivered to any part of the world, at relatively low cost and with no time lag. Thus US judicial decisions are abstracted and entered into an electronic database in Korea; telephone directory enquiries in the United States are supplied from the Caribbean.

In banking, information technology has made it possible for institutions to centralize information resources in areas such as derivatives trading and market forecasting on a global basis. These same facilities can be used to offer services to clients, as when a bank sells access to its global information system, enabling corporate treasurers to monitor balances and financial data around the world. Internal managerial and support services can be exported and imported from parent to subsidiary, or from subsidiary to subsidiary. Back-room operations can be shifted overseas to lower-cost environs or domestic locations remote from the major financial centres. Firms can engage in 'outsourcing' of products and processes.

Competition among centres for this business has been both across and within borders. Clearing facilities for international bonds have been attracted to Brussels and Luxembourg. Special tax exemptions have played a part in the promotion of Dublin's International Financial Services Centre as a location for activities such as global custody and treasury operations. But a whole host of other factors are important, depending upon the nature of the activity and the type of 'back office' and supportive system associated with the financial service. Table 1.1 classifies seven types of financial services provided by banks: deposit and payments services; consumer loans; commercial loans; merchant banking; investment management; risk management services; and services related to international trade. For each service, the table identifies the major products involved, and then the nature of the delivery systems for them, and the associated back-office systems and supportive services. Consider, for example, the case of consumer loans in the form of home mortgages, personal loans and personal overdrafts. Delivery can take place via branch lending officers, via telephone banking or through various agency arrangements with other financial institutions (e.g. insurance companies) and enterprises (e.g. real estate companies). Provision of the loans relies on a variety of 'back-office' systems such as access to a customer database, credit scoring, credit control, processing, servicing, preparation of

Table 1.1 Classification of banking services and associated systems

Nature of service	Products	'Front-office' (delivery) systems	'Back-office' (productive) systems	Supportive systems
Deposits and payments services	Cheque book accounts Savings accounts Investment accounts Term deposits Credit cards Standing orders Direct debits/credits Automated requests Giro transfers	Counter staff Lobby terminals ATMs Telephone banking Home banking Video link Supermarket branching Internet banking	Transactions processing systems Credit card accounting Cheque card fraud monitoring Ordering cheque books Balance enquiries Calculating charges and interest Account records and passbook entries	Clearing houses CHAPS, BACS Visa/Mastercard networks ATM networking
Consumer loans	Home mortgages Personal loans Personal overdrafts	Branch lending officers Telephone banking Agency arrangements	Loan and mortgage processing Credit scoring and control procedures Loan and mortgage servicing Mortgage repayment tables APR quotations Early repayment quotations Customer data base	Real estate evaluation Conveyancing Lodgement of security Endowment insurance services Loan packaging and securitization services
Commercial loans	Overdrafts Term loans Leasing Factoring Credit lines and facilities Management buyouts Venture capital Project financing	Corporate accounts officer Corporate Banking Centres Small-business offices Mobile lenders	Credit assessment and credit control Storage of accounting and business information Structuring of loan covenants	Legal services Accounting services Loan syndicates

Merchant banking	Securities underwriting Securities distribution Private placements Broking/market-making of securities Mergers and acquisitions	Client liaison officers Institutional sales teams Telephone, telex, fax and other communications	Market intelligence systems Dealing systems Settlement systems	Underwriting syndicates Distribution and selling institutions News and information services CEDEL, Euroclear
Investment management services	Portfolio management Unit trust management Pension fund management Insurance underwriting Tax and financial advisory services Custodial services	Institutional and business liaison officers Management teams	Processing and registration of securities transactions Monitoring market developments Calculating investment strategies Valuing asset portfolios Preparing accounting reports	Accounting and legal services Property and valuation services News and information services Subcustodians and reinsurers
Risk management services	Options Futures Forwards Swaps Caps, collars Energy and commodity futures Risk-planning advisory services	Corporate accounts officers Client liaison officers and sales teams	Software for pricing options, swaps, etc. Calculation of net interest and currency payments Monitoring risk exposures and margining requirements	News and information services Accounting and legal services
International trade services	Foreign exchange and funds transfers Forward cover Documentary letters of credit International trade advice and intelligence Forfaiting	Branch manager Accounts officer International banking centre	Processing of international funds transfers Processing of documentation	Correspondent links Overseas branch networks

ATM, automated telling machine; APR, annual percentage rate; CHAPS, clearing house automated payment system; BACS, bankers' automated clearing services; CEDEL, Centrale de Livraison de Valeurs Mobilières

repayment tables and annual percentage rate (APR) quotations, determination and pricing of early repayment risk. Supportive systems required for lending include real estate valuation, conveyancing, lodgement of security, endowment insurance and, should the loans be on-sold, loan packaging and securitization.

Some of these supportive and back-room requirements can be 'outsourced' from other financial enterprises which themselves specialize in this type of

Table 1.2   Determinants of the location of ancillary financial services activities

Space requirements	Availability of greenfield sites Land prices Property prices Office rents Quality of office space
Labour pool	Supply of labour Male/female mix Labour turnover Cost of labour
Labour skills	Data-processing skills Keyboard skills Facility with computers Graduate recruitment
Communications	Advanced telecommunications Fibre optic, satellite links Road transport links Rail, high-speed trains International/domestic airlines
Technological facilities	Computer hardware Software systems Automated delivery systems Head office and branch facilities Networking capacity Access to database
Complementary services	Accounting and tax advice Legal services Management services Real estate, conveyancing Property management
Systems security	Bank robbery Terrorism Labour/union disruption
Quality of life	Standard of housing Cost of housing Education Cultural activities Recreational facilities

activity. But in all cases, it is apparent that there is considerable potential to shift or locate many of these systems in places away from the major financial centres. Table 1.2 lists some of the determinants of these location decisions. Considering the case of the United Kingdom, it is often overlooked that despite the dominance of the City, London accounts for less than 30 per cent of employment in the financial services industry. In the ten years to July 1994, Jones Lang Wootton records financial activities occupying a total of 5.8 million square feet (540,000 m^2) of office space as having been relocated from central London to outer London and the provinces. Space requirements and lower office costs have been major factors in the expansion of Bristol, Cardiff and Leeds as locations for financial services, as was ease of travel to London. Availability of staff with good skills, and the close proximity of advertising, legal and printing services have been other factors. Thus the insurer Norwich Union chose Dublin as the place for its computing centre because of the difficulty of recruiting computing staff in the UK. Skilled workers and professionals also have pronounced living preferences, which favour Edinburgh over Glasgow, Dublin over Belfast, Leeds over Sheffield.

Leeds is one example of an ancillary centre. Employment in financial services in Leeds grew from 27,500 to 45,700 in the ten years from 1981, and the financial sector now provides work for about 1 in 10 of the city's population. As well as being a headquarters for building societies, the Leeds region is now the centre for telephone-delivered financial services by Direct Line Insurance and First Direct. GE Capital Retailer Financial Services and Club 24, both of which handle credit card accounts, are also headquartered in Leeds. Based around such developments there are accountants, solicitors, computer consultants, security printers, venture capitalists, treasury officials and other professionals associated with financial services. In 1994, the Leeds Financial Services Initiative sought to build on this growth and promote Leeds as the second city in the United Kingdom for financial services, with the specific aim of persuading foreign (particularly European) and national merchant banks and financial institutions which are contemplating relocation to consider Leeds.

Comparing the Centres

A typology of European financial centres under the single European financial market has been suggested by Begg (1991). This classification is set out in Table 1.3 and ranges from global centres down to specialist national centres and smaller retail outlets. Being concerned with the European Union, the table neglects the role of offshore locations such as Jersey and the Isle of Man, as well as financial centres such as Switzerland.

Because of scale economies and the gravitational pull of concentrated trading, only one truly global centre may be able to exist within Europe, with London clearly at present number one (Abraham *et al.*, 1994). But as diversi-

Table 1.3   Typology of European financial centres for the single market

Type of centre	Characteristics	Cities in the category
Node of global financial system	Diversified services; good communications; HQs of major high-volume securities and markets; branches of foreign banks, etc.	London and Paris; Frankfurt to a lesser extent
Diversified European centres	Significant cross-border business; HQs of major financial and non-financial companies; critical mass in markets; air and telecoms links; location of regulatory organizations; wide range of complementary business services	London, Paris, Frankfurt; scope for Amsterdam, Milan, Madrid and possibly Brussels
Specialist European centres	Strong in niche markets; critical mass of specific labour skills	Edinburgh, Luxembourg; Dublin gaining ground
National centres with limited international business	Presence of several leading financial entities; complementary services; mix of general and specialist services, some financial markets and regulatory functions	Amsterdam, Brussels, Rome, Milan, Madrid, Athens, Lyons, Hamburg, Lisbon, Munich, Barcelona, Dublin
Regional centre	Wide range of regional offices; mixture of business services; some financial HQs; back offices	Düsseldorf, Hanover, Berlin, Bordeaux, Lille, Marseilles, Rotterdam, Eindhoven, Turin, Bologna, Palermo, Naples, Liège, Antwerp, Salonika, Oporto, Bilbao, Valencia, Málaga, Bristol, Leeds, Glasgow, Liverpool, Belfast, Birmingham
Specialist national centres	Centre for processing activity or presence of a major financial entity	Halifax, Utrecht, Norwich, Mons, for example
Retail outlet	Predominantly retail distribution of financial services	Other cities and towns

Source: Raikes and Newton (1994). Reproduced with kind permission from Kluwer Academic Publishers

fied European centres – Begg's second category – London, Frankfurt and Paris are more evenly matched, particularly if the second two form the 'inner core' of a monetary union. Some of the strongest competition, however, seems likely within the fourth category, that of centres which are presently oriented mainly to domestic markets, but harbour ambitions to gain some share of international business.

Considering the third category, that of niche or specialist centres, the existence and continued success of Edinburgh, Luxembourg and, latterly, Dublin seem to cut across the centripetal forces at work in financial markets. These centripetal forces favour having payments flows, seasonal surpluses and deficits, and savings and investments all routed through a single centre, so achieving economies in the number of transactions. A central marketplace for trading enables speed of interaction and eliminates the need to search over wide distances to discover price information and eliminates the need to search over wide distances to discover price information and to maintain continuous interchange, while allowing economies of scale in brokerage, handling and settlement systems. Markets attract around them support facilities in the form of rating and news agencies, communication facilities, and accountants, lawyers, securities dealers and valuers. The process is self-reinforcing. A broader market makes for liquidity and encourages more participants to transfer dealing to the central venue, so enhancing liquidity. Greater trading makes investments in market support systems worthwhile, which encourages more trading.

Yet there are factors which work against centralization and which favour local operations. Foremost among these are the costs of collecting information, localized share ownership, regional preferences in investing, knowledge of local identities, and a regional ethos. These give local centres some advantages. Luxembourg and Dublin, for example, have achieved strong growth in the development of trust funds and other investment products. Although both centres have relied on tax advantages for their expansion, they are also examples of what we have termed *ancillary centres*, where the servicing and administration takes place of funds managed in the larger centres or where there is wholesaling (as in fund management) on behalf of other suppliers elsewhere.

The remainder of this section compares the centres market by market, although it must be noted that the data – like the markets themselves – do not cover Europe only, and are usually on a world basis.

International banking

Table 1.4 shows the composition of international banking by location and type of transaction as at March 1994. Under the Bank for International Settlements (BIS) definition, international banking includes cross-border positions in domestic and foreign (Euro) currencies along with local (cross-currency) lending in foreign currency. The United Kingdom (almost entirely London) has regained the position of largest centre, which it lost to Japan in the early 1990s, but continues to decline in size relative to other European and the offshore centres. The most notable feature of the table is the continued

Table 1.4   International bank lending, March 1994 ($ billion)

Reporting centre	Traditional foreign lending	Cross-border Eurolending	Foreign currency lending to residents	Total lending	%
Austria	14.8	49.7	18.7	83.2	1.06
Belgium	23.7	186.2	58.9	268.8	3.44
Luxembourg	7.6	363.5	66.3	437.4	5.60
France	144.3	359.6	82.8	586.7	7.51
Germany	246.2	184.2	15.7	446.1	5.71
Italy	59.0	81.5	99.3	239.8	3.07
Netherlands	31.5	148.3	36.8	216.6	2.77
Switzerland	92.9	284.9	21.7	399.5	5.11
United Kingdom	88.8	983.0	294.1	1,365.9	17.50
Other European	64.7	154.6	84.3	303.6	3.89
Sub-total: All European countries	773.5	2,795.2	778.6	4,347.3	55.66
Canada	4.1	42.8	27.0	73.9	0.95
Japan (JOM)	174.7	808.0 (549.9)	344.5 (50.2)	1,327.2	17.00
United States (IBF)	283.3	263.0 (246.1)	—	546.3	6.99
Caribbean centres	—	573.7	28.3	602.0	7.70
Asian centres	13.4	896.8	3.8	914.0	11.70
All reporting centres	1,249.0	5,379.5	1,182.3	7,810.7	100
%	15.99	68.87	15.14	100	

Source: Bank for International Settlements (1994). *International Banking and Financial Market Developments*, August
Note: JOM = Japan offshore market; IBF = international banking facility

decline of New York, reflecting in part the move of US banks into off-balance sheet business.

Foreign exchange

London is by far the largest base for foreign exchange trading, with a daily turnover in April 1995 six times larger than that of Germany and a market share equal to 30 per cent among the 26 countries surveyed. This is more than one and a half times the turnover recorded in 1992 and more than five times the turnover of 1986 (see Table 1.5). While the United States obviously remains dominant in the American time zone, foreign exchange turnover in Japan is now well below that in Singapore and Hong Kong combined. In Europe, small countries such as Switzerland, Denmark and Luxembourg have a strong market position.

Table 1.5 Foreign exchange market activity by country (average daily net turnover in US$ billion)

Country	March 1986[1]	April 1989	April 1992	April 1995
United Kingdom	90	184	291	465
United States	58	115	167	244
Japan	48	111	120	161
Singapore		55	74	105
Hong Kong		49	61	90
Switzerland		57	66	87
France		23	35	58
Germany		n.a.	55	76
Australia		29	29	40
Denmark		13	27	31
Canada		15	22	31
Belgium		11	16	28
Netherlands		13	20	26
Italy		10	16	23
Sweden		13	21	20
Luxembourg		n.a.	13	19
Spain		5	12	18
Austria		n.a.	5	13
Norway		4	5	8
New Zealand		n.a.	4	7
Finland		4	7	5
South Africa		n.a.	4	5
Ireland		5	6	5
Greece		1	1	4
Bahrain		3	4	3
Portugal		1	1	3

Source: Bank of England, Bank for International Settlements
[1]Survey confined to these countries

Securities markets

Four comparisons of equity markets are shown in Table 1.6, namely, the listing of foreign companies, market valuation, total turnover and liquidity (turnover relative to market value). Listing by foreign companies is one indicator of the internationalization of a centre. On this measure, Amsterdam, Brussels, Germany, Luxembourg, Paris, London and Switzerland can be said to be international equity markets, while Copenhagen, Helsinki, Madrid, Milan, Oslo and Stockholm are not. The US and Japanese exchanges also rely on domestic companies for their importance. Turnover of the foreign equities listed is another measure, and the data for 1995 indicate that London has about 60 per cent of the world market in this area.

Table 1.7 provides another perspective on the relative importance of the European centres for securities business. It shows the location of member firms of the International Securities Market Association. London, Switzerland and Luxembourg in that order have the most Association members. In terms of the location of market-making members, the position is a little

Table 1.6   International stock-market comparisons, 1995

Exchange	Number of listed companies at year end		Market value of listed companies at year end	Turnover value			Liquidity turnover/ market value
	Domestic	Foreign	($USm)	Domestic ($USm)	Foreign ($USm)	Total ($USm)	(%)
Europe							
Amsterdam	217	216	284,312	122,088	189	122,277	43.0
Athens	197	—	16,347	5,941	—	5,941	36.3
Brussels	143	138	100,516	15,049	3,072	18,121	18.0
Copenhagen	213	10	55,669	26,802	841	27,643	49.7
Dublin	62	9	25,351	6,918	—	6,918	27.3
Germany	437	220	574,939	588,465	14,456	602,921	104.9
Helsinki	73	—	43,373	18,657	—	18,657	43.0
Lisbon	169	—	18,192	4,113	—	4,113	22.6
London	2,303	536	1,396,432	497,676	608,904	1,106,580	79.2
Luxembourg	55	228	29,397	459	14	473	1.6
Madrid	361	4	188,896	50,663	14	50,677	26.8
Milan	217	4	203,711	92,412	32	92,444	45.4
Oslo	151	14	44,153	24,029	314	24,343	55.1
Paris	710	194	494,719	204,147	3,533	207,680	42.0
Stockholm	212	11	170,527	84,449	955	85,404	50.1
Switzerland	216	233	394,851	296,128	15,911	312,039	79.0
Vienna	95	38	30,390	12,705	240	12,945	42.6
Others							
American	791	n.a.	136,167	72,132	n.a.	72,132	53.0
Australian	1,129	49	244,386	97,864	770	98,634	40.4
Hong Kong	518	24	303,870	103,543	240	103,783	34.2
Johannesburg	614	26	278,253	16,135	1,075	17,210	6.2
Korea	721	—	178,175	180,400	—	180,400	101.2
Mexico	186	—	90,207	35,206	—	35,206	39.0
Nasdaq	4,760	362	1,150,600	2,279,283	99,621	2,378,904	206.8
New York	2,428	247	5,719,699	2,803,157	259,870	3,063,027	53.6
New Zealand	145	60	31,788	n.a.	n.a.	8,448	26.6
Singapore	258	36	150,943	n.a.	n.a.	—	0.0
Taiwan	347	—	185,849	369,321	—	369,321	198.7
Tokyo	1,714	77	3,516,655	803,534	939	804,474	22.9
Toronto	1,196	62	365,684	151,554	430	151,984	41.6

Source: London Stock Exchange Fact Book 1996
Note: Nasdaq = National Association of Securities Dealers Automated Quotation System

different. Nearly two-thirds of ISMA reporting dealers are located in London. This measure accords with other indicators of international securities activity. Recent estimates (Raikes and Newton, 1994) suggest that around 60 per cent of primary international (Euro) bond syndication takes place in London and that 75 per cent of secondary market trading of Eurobonds is centred in London.

Table 1.7 Location of International Securities Market Association member firms (number of firms by centre, January 1994)

	All firms	Reporting dealers
United Kingdom	172	69
Switzerland	130	2
Luxembourg	65	5
Germany	64	2
Italy	54	9
Netherlands	47	1
France	42	11
Hong Kong	36	1
Belgium	35	2
United States	29	1
Other	198	7
Total	**872**	**110**

Source: ISMA, reported in Raikes and Newton (1994). Reproduced with kind permission from Kluwer Academic Publishers

Derivatives markets

Turnover in exchange-traded derivatives is listed in Table 1.8, which shows annual contracts traded in the largest futures and options exchanges in 1992 and 1993. While the two Chicago exchanges still dominate world trading, their relative position has declined *vis-à-vis* the European markets, particularly LIFFE, MATIF and DTB, all of which have recorded strong growth in recent years. Trading in the over-the-counter (OTC) derivatives has grown more rapidly than the exchange-traded instruments, and the OTC markets now exceed the exchanges in size. The latest BIS data show that at the end of 1995, amounts outstanding of OTC contracts totalled $18.0 trillion as compared with $9.2 trillion for the exchange-traded financial instruments (Bank for International Settlements, 1996). New York and London are the market leaders for this business, followed by Tokyo and the continental European centres.

Fund management

Growth of the fund management industry has been stimulated by a number of developments. Following the abolition of exchange controls in a large number of countries, institutional investments have diversified to include foreign holdings. With population ageing and the expansion of private occupational superannuation schemes, more funds have been directed to the long end of the capital market. The result is that fund managers now account for a major portion of trading in currencies and securities, and their decisions about where to do business play an increasing role in the location of finance.

Table 1.8   Futures and options exchanges: annual turnover

Exchange		Number of contracts traded	
		1993	1992
Chicago Board of Trade	CBOT	178,773,105	150,030,460
Chicago Mercantile Exchange	CME	146,746,990	134,238,555
London International Financial Futures Exchange	LIFFE	97,108,712	71,977,025
Marché à Terme International de France	MATIF	72,263,961	55,474,238
New York Mercantile Exchange	NYMEX	55,412,436	47,212,417
Bolsa de Mercadorias e Futuros, Brazil	BM&F	52,263,359	35,072,146
Deutsche Terminbörse	DTB	37,924,127	34,842,778
Tokyo International Financial Futures Exchange	TIFFE	24,126,147	15,540,487
Tokyo Commodity Exchange	TOCOM	21,557,795	13,585,379
Sydney Futures Exchange	SFE	21,481,096	17,557,685
Tokyo Stock Exchange	TSE	18,979,492	14,538,717
Commodity Exchange Inc.	COMEX	18,854,113	12,673,179
OM Stockholm AB	OMS	16,044,412	17,147,096
Singapore International Monetary Exchange	SIMEX	15,729,787	12,180,174
Osaka Securities Exchange	OSE	14,551,833	21,184,310
International Petroleum Exchange of London	IPE	13,769,978	10,674,803
Tokyo Grain Exchange*	TGE	13,687,746	12,416,671
Coffee, Sugar and Cocoa Exchange	CSCE	11,304,823	9,275,708
Swiss Options and Financial Futures Exchange	SOFFEX	6,808,693	9,258,859

Source: Futures Industry Association
*Exchange merged with Tokyo Sugar Exchange on 1 October 1993

Fund management is an activity for which only sparse data exist: the information in Table 1.9 is patchy and relates only to equities. It is also an operation which encapsulates the complexity of the locational decision in financial services. Fund managers have to balance physical proximity to customers, closeness to the markets in which they deal and relations with other fund managers. Portfolio management can be separated from administration and other 'back-office' activities, and the type of fund is also a conditioning factor. The result is the existence of significant centres of fund management outside the major financial centres. In the United States, Boston had about 10 per cent of equities under management in 1990 as compared with 27 per cent in New York. In Germany, Frankfurt is the main centre for fund management, but there is a subsidiary centre in Munich, which is about one-third of its size. In Switzerland, most of the fund management is divided between Zurich and Geneva, although Basle has a small business. Zurich primarily serves institutional clients and Geneva serves private clients. The consequence of this specialization has seen Zurich expand relative to Geneva,

Table 1.9   Institutional equity holdings in six European centres

	London (%)	Paris (%)	Frankfurt (%)	Edinburgh (%)	Zurich (%)	Geneva (%)	Total ($m)
1987	32.0	6.7	6.0	4.9	24.4	26.0	750
1988	32.7	6.8	5.8	4.7	24.6	25.4	860
1989	32.5	7.2	6.5	5.6	22.9	25.2	1,123
1990	32.6	11.3	9.7	5.5	19.7	21.2	1,256
1991	35.1	11.8	10.6	5.8	18.1	18.7	1,358
1992	35.4	12.5	10.1	5.8	22.1	14.1	1,505

Source: Kay *et al.* (1994)

along with the growth of institutional investment. Edinburgh has carved out a niche in this industry, and this is considered further in the next section.

Developments in Particular Locations

This section examines some of the competitive forces at work among European centres by considering developments in particular centres.

Frankfurt

'Finanzplatz Deutschland' is not the result of any official interventions; the initiative has rested almost exclusively with the enterprises and associations in the financial markets. This has produced a series of gradual steps, with facilitating legislation to amend controlling laws. There has been no 'big bang' as in London and, later, Paris. Thus the advent of electronic trading in the form of the screen-based Integrated Stock Exchange System (Integriertes Börsenhandelssystem – IBIS) and the computer-based order matching on the German Financial Futures Exchange (Deutsche Terminbörse – DTB) came from market innovations. The integration of the various stock exchanges into the German Stock Exchange (Deutsche Börse AG) came without the prompting of public bodies, which restricted themselves to defining underlying legal requirements. A rapid expansion of trading has resulted from these moves, enhancing Frankfurt's position as an international and European centre, which many expect to be reinforced by the location of the European Monetary Institute – and perhaps the future European Central Bank – in Frankfurt.

In the past, Frankfurt's role as a financial centre has been impeded by two factors: regulations, and competing claims from other German cities. Heavy-handed regulation of banking saw German banks establish subsidiaries in Luxembourg to avoid reserve requirements and other government controls over domestic operations. Queuing and other constraints upon issues of securities in Germany started a cumulative process which has resulted in leading German banks shifting their international securities operations to

London. A substantial lowering of minimum reserve requirements in 1993 and 1994 along with changes in the domestic securities markets – including from August 1994 a lifting of the ban against the authorization of money market funds – has been seen as evidence of a change of heart by officialdom. Nevertheless, the traditional conflict remains. For the Bundesbank, the safeguarding of the currency has clear priority over promoting Germany as a financial centre (Häusler, 1994), and it is prepared to wear the costs of that overriding mandate. As a consequence, the Bundesbank retains a non-interest-bearing reserve ratio requirement – albeit at a reduced level – to stabilize the demand for central bank money as a fulcrum for monetary control. Also for monetary control purposes, there continues to be a prohibition upon the issue abroad of DM certificates of deposits.

Frankfurt has also had to contend with strong regional centres in Munich and Hamburg – perhaps Berlin in the future – although with the launch of the 'Finanzplatz Deutschland' campaign there has been greater acceptance of Frankfurt's claim for prominence. However, the federal and regional structure of Germany still hinders Frankfurt. As was emphasized above, financial services rely on a considerable supporting infrastructure, and a financial centre has to offer a wide range of characteristics to attract business away from other locations. London and Paris are the centres for *everything*, allowing ready contact with the heads of multinationals, politicians, legislators and administrators, heads of international bodies, regulators, trade associations, professional institutes, etc. London is the headquarters for 86 of the top 100 UK manufacturing corporations. In France, Paris is the head office for 78 per cent of the leading 500 French firms. In Germany, Bonn (soon to be Berlin) is the political centre. Hamburg is the centre for publishing, communications and advertising. Munich and Hamburg are important in insurance. Germany also has a dispersed management structure in industry with Hamburg, Frankfurt and Düsseldorf as the most important bases, closely followed by Essen, Cologne and Hanover.

Switzerland

The old saying 'Money alone can't make you happy, unless it's in Switzerland' summarizes Switzerland's reputation for looking after other people's money. The data confirm this foreign orientation to be still the case. Foreign liabilities represented 30 per cent of the balance sheet totals of Swiss banks in 1991. At the same time, non-resident liabilities constituted 76 per cent of the fiduciary accounts of the Swiss banks (Blattner, 1994).

Bank secrecy laws are one attraction of Swiss banks and an example of how the reality differs from the image. Switzerland has been forced to bow to international (mainly US) pressure, and banks may reveal information relating to their account holders to foreign governments in criminal prosecutions. Notably, also, the Berne government blocked the deposits of ex-President

Marcos to stop him transferring them out of the country. Secrecy laws are now stricter in many other countries; for example, violation of banking secrecy is a criminal offence in Austria.

Clearly, people choose to invest in Switzerland for many reasons entirely unrelated to secrecy laws: the range of services on offer from the universal banks; low inflation; freedom from exchange controls and other restrictions; political stability; and the reputation of Swiss banks for safety and soundness. Yet these are all areas in which Switzerland's traditional competitive advantages have been eroded. Universal banking operates in many centres, and the universal banking model has been adopted as the norm for the single European market. Many other countries now offer the prospect of low and stable inflation, along with political stability. Exchange controls are now a rarity and most other financial markets have been liberalized and modernized; indeed, in some areas Switzerland has been left behind, and has had to catch up by integrating its three stock exchanges and by introducing screen-based computerized trading. Finally, Swiss banks have not been immune from the real estate boom and bust, with some regional banks experiencing severe difficulties (one collapsed). Also, since the institution of the BIS capital adequacy standards, Swiss banks are no longer so well-capitalized as compared with other banks. When to these diminished advantages are added the traditional disadvantages of Switzerland, namely a small market and the propensity to tax securities business by means of stamp duties and withholding taxes, it is easy to understand why there is something of a crisis of confidence in Swiss markets.

Edinburgh

Although the great majority of UK fund management is undertaken in London, there are firms located elsewhere. They fall mainly into two categories: those located in the similar but smaller centre in Scotland, and large regional insurance companies such as the Norwich Union and Cooperative (Manchester) with their own retail client bases. There are also some smaller firms established in other locations, such as Abbey Life and MFMG in the Bournemouth/Poole area.

About 14 per cent of total UK funds are managed in Scotland. The dominant groups are institutional investors and fund managers. Institutional investors are pension funds, life insurance companies and investment trusts, while the fund managers are institutions that manage portfolios for others. These two groups overlap because pension funds are prominent among the clients of fund managers, and institutional investors sometimes subcontract part of their portfolios to independent fund managers. Also, trust management companies themselves, along with life insurance companies, often act as fund managers for other institutions. This business is international; fund managers in London and Edinburgh manage large sums for US pension funds (Revell, 1994).

The fund management centre in Edinburgh is the fourth largest in Europe, with three clearing banks, six major life insurance companies, nine investment managers and one independent investment trust. Funds under management in Edinburgh in 1992 totalled at least £25 billion, and 830 people were employed in fund management. Total funds under management in Scotland came to over £100 billion in 1992, of which investment managers had £32 billion, and employed 1,350 people. Scottish life offices had £72 billion under management (Kay *et al.*, 1994).

A balancing of factors favours location in Scotland. Staff costs per head are around 20–25 per cent lower than in London and the south of England, and accommodation costs per square foot are over 50 per cent lower. Against these must be set efficiency losses, for Edinburgh fund managers have less to manage – around £30m per employee against about £53m in the rest of the United Kingdom. Perhaps for this reason, the fund managers in Edinburgh tend to concentrate on a narrower range of products than those in London in order to keep down costs. Nevertheless, it is apparent from Table 1.9 that Edinburgh has increased its share of institutional equity investments.

British offshore centres

Jersey, Guernsey and the Isle of Man are UK Crown dependencies with their own governments which have the right to set their own income tax levels. All three levy tax at 20 per cent. Individuals can use offshore facilities to form companies and trusts to hold and manage assets, or avoid inheritance or capital gains tax (Doggart, 1993). They can put money into banks, investment companies, building societies and life assurance companies to defer tax payments on interest. Some with high incomes choose to take up residence offshore. Companies can greatly reduce costs by using offshore centres to set up captive insurance companies, treasury centres and subsidiaries, or to register ships.

Although most services are available in all three centres, each has tended to develop its own speciality. Jersey led the way into offshore finance with a rush of trusts moving offshore in the early 1970s, following a change in UK law. The island has continued to build on this, and its 1984 trust law provided statutory backing. Currently it is probably best known for its trust and private banking business, but fund management is increasing rapidly. Funds under management in Jersey now exceed £25 billion, serving 150,000 investors from a wide range of geographical areas. There is a greater choice of funds offshore – derivatives, overseas funds and types not found in the United Kingdom – and there may be tax advantages for the investor, particularly for registered non-taxpayers (Hampton, 1994).

Neighbouring Guernsey now ranks third in the world for its number of offshore captive insurance companies, currently around 250, behind Bermuda (1,300) and the Cayman Islands (390). (A captive is a wholly owned subsidiary of its parent company and insures all or some of its parent's risks.) Guernsey moved into the market in the 1970s and has built up a large pool of expertise in this area. Premiums are deductible for tax purposes, which gives an added cost advantage.

The Isle of Man has around 150 captives, but its main growth area has been in offshore life insurance. The island has a policyholder's protection scheme which covers 90 per cent of the amount invested in the event of a company failure. Offshore policies offer maximum freedom of choice in investments and switching investments. Several major life companies are based in the Isle of Man, and their clients mainly fall into three categories: those living and working overseas who have a lump sum to invest and require a single-premium policy; those working overseas who make regular savings through a policy; and anyone else who wishes to invest in an environment where funds accumulate tax-free.

However, the days are long gone when just about anyone's cash was welcome. Offshore banks, institutions and professionals are governed today by a strict 'know your customer' policy. For this reason referrals from professional advisers are the most common form of introduction. There is little in the way of a veil of secrecy for transactions. Nevertheless, despite tightened regulatory standards, scandals still occur, such as the collapsed foreign exchange fund in Jersey in 1996.

Dublin

Dublin's International Financial Services Centre (IFSC) is the most recent addition to the offshore centres, although the IFSC prefers to be labelled 'onshore' low tax rather than offshore. Like International Banking Facilities (IBFs), it is an example of 'onshore offshore' finance (Lewis and Davis, 1987). The IFSC illustrates three features of financial centres. First, there must be some natural assets or endowments. Second, to these can be grafted some *created* assets. Third, there is a role for smaller centres looking to particular market niches.

In Ireland's case, it possesses a stable legal and regulatory environment and modern telecommunications. It is in the same time zone as London and is English-speaking. There is an ample supply of young, well-educated workers, available at relatively low labour costs, some of whom had moved to London for work and experience, and wished to return home. To these extant assets was added a reclaimed development site in a central, riverside location with a potential for 1 million square feet (93,000 m^2) of office space and a range of tax incentives to locate there. Chief among these is the corporation tax: companies operating in the IFSC in activities carried out with non-residents in currencies other than the Irish punt are subject to a 10 per cent corporate tax rate, approved by the EU to operate until 2005. The other tax exemptions are zero local tax, double deduction of rent from taxable income, a 100 per cent capital allowance in the first year of trading, 100 per cent write-off of new equipment expenditure, capital allowances for leasing, absence of capital gains tax on traded income and absence of value-added tax on output.

Table 1.10 gives details of the 221 IFSC companies approved and active as at the beginning of 1994. The direct tax receipts from IFSC companies in 1993 totalled IR£133 million, although there may have been some displacement by Irish firms of pre-existing foreign-currency business with non-residents into the lower-tax environment of the IFSC. In the certification process, priority is given to employment potential and local services, rejecting 'brass-plate' operations and also those where the business is effectively done elsewhere, leaving only a low-skill component in the IFSC to avoid taxation in another country. So far the employment commitment from the 221 companies is 2,800, with about half of the jobs now in place. This is well below the forecast employment of 5,000 for 1992 (O'hUiginn, 1990) and the projected total of 7,500 jobs when complete (*Euromoney*, 1988).

Insurance, treasury management and collective investments are the three main business areas attracted to the IFSC, and these account for almost half of total employment (O'Connell and Kennedy, 1994). The IFSC has become a popular venue for insurance, with 22 insurance managers and 90 captive insurance companies. Dublin's competitive advantage over other offshore locations such as Guernsey and the Isle of Man comes from the ability to write insurance directly into EU countries, for, unlike these other centres, Ireland is a full member of the EU. Treasury management in the centre

Table 1.10   Number of International Financial Services Centre projects classified by region of origin and business category, spring 1994

Business	Country						
	Europe	Ireland	Canada	USA	Japan	Other	Total
Insurance	19	2	1	—	—	5	27
Group treasury/asset financing	58	16	10	19	2	6	111
Moneybrokers	—	2	—	—	—	—	2
Bank affiliate/ subsidiaries	13	7	—	4	—	1	25
Collective investment-type project	5	—	—	6	—	—	11
Other IFSC projects	24	8	2	2	4	5	45
Total	119	35	13	31	6	17	221

Source: O'Connell and Kennedy (1994). Reproduced with kind permission from Kluwer Academic Publishers

Table 1.11 Collective investment funds in various centres (IR£ billion equivalents)

	1992	1993
Dublin	3.7	8.9
Jersey	16.8	24.9
Guernsey	8.8	12.8
Isle of Man	2.7	5.9
Luxembourg	n.a.	106.0

Source: O'Connell and Kennedy (1994). Reproduced with kind permission from Kluwer Academic Publishers

specializes in corporate treasury operations for multinational companies. For example, Wang International carries out foreign exchange hedging in the IFSC, along with factoring, pension fund management and insurance. The corporate tax rate of 10 per cent is much lower than that of alternative locations such as Luxembourg, Belgium and the Netherlands, but the IFSC is disadvantaged relative to the latter two in terms of a lack of double-taxation agreements. In terms of collective investments, Table 1.11 shows that Dublin has already passed the Isle of Man, and has 146 schemes registered, comprising 376 funds. Total funds under management were IR£8.9 billion as at the end of 1993.

Thus although the IFSC has not come up to the undoubtedly over-optimistic projections made for its growth, it must be remembered that not many years have elapsed since the idea was first mooted. Since then the IFSC has developed steadily and, notwithstanding large gaps on the site itself, seems to have established a viable entity around a number of niche activities.

Conclusion

The message of this chapter is simple. Provision of financial services is a complex operation embracing a large number of elements. These elements can be grouped into a number of core components; this chapter considers six. It has been shown that market and technological developments over the past three decades have produced a 'splintering' and geographical dispersion of these components. As a consequence, a large variety of factors interact to influence the location of financial services activity.

Abraham *et al.* (1994) in fact classified no fewer than 47 factors in the location decision. However, their focus was with *international* financial centres. Most of the European financial centres also have a strong regional orientation, and the examination here of particular markets and centres, such as Edinburgh, Dublin and the British offshore islands, shows there to be keen competition within the time zone for the provision of specialized services and the complementary and ancillary activities.

References

Abraham, J.-P., Bervaes, N. and Guinotte, A. (1994) 'The Competitiveness of European International Financial Centres.' In J. Revell (ed.) *The Changing Face of European Banks and Securities Markets*. London: Macmillan, pp. 229–284.

Bank for International Settlements (1996) *66th Annual Report*, 10 June, Basle.

The Banker (1992) 'Mediterranean Banking: A Place in the Sun', vol. 142, October: pp. 48–50.

Begg, I. (1991) 'The Spatial Impact of Completion of the EC Internal Market for Financial Services.' *Journal of Regional Studies*, vol. 26, no. 4: pp. 333–347.

Bhagwati, J.N. (1984) 'Splintering and Disembodiment of Services and Developing Nations.' *The World Economy*, June, vol. 7, no. 2: pp. 133–143.

Blattner, N. (1994) 'The Swiss Financial Centre Revisited.' In D.E. Fair and R. Raymond (eds) *The Competitiveness of Financial Institutions and Centres in Europe*. Dordrecht: Kluwer Academic Publishers.

Doggart, C. (1993) *Tax Havens and Their Uses*, revised edition. London: Economist Intelligence Unit.

Dufey, G. and Giddy, I. (1978) *The International Money Market*. Englewood Cliffs, NJ: Prentice-Hall.

Euromoney (1988) 'Dublin, the New Financial Centre.' Sponsored supplement, *Euromoney*, March.

Hampton, M. (1994) 'Treasure Islands or Fool's Gold: Can and Should Small Island Economies Copy Jersey?' *World Development*, vol. 22, no. 2: pp. 237–250.

Häusler, G. (1994) 'The Competitive Position of Germany as a Financial Centre as Seen by a Central Banker.' In D.E. Fair and R. Raymond (eds) *The Competitiveness of Financial Institutions and Centres in Europe*. Dordrecht: Kluwer Academic Publishers.

Johns, R.A. (1992) 'Offshore Banking.' *New Palgrave Dictionary of Money and Finance*, vol. 3. London: Macmillan.

Kay, J., Laslett, R. and Duffy, N. (1994) *The Competitive Advantage of the Fund Management Industry in the City of London*. City Research Project, Subject Report IX. Corporation of London, February.

Kindleberger, C.P. (1974) *The Formation of Financial Centers: a Study in Comparative Economic History*. Princeton Studies in International Finance, no. 36. Princeton, NJ: Princeton University Press.

Leveson, I.F. (1982) *The Economic Future of the United States*. Boulder, CO: Westview Press.

Lewis, M.K. and Davis, K.T. (1987) *Domestic and International Banking*. Deddington, Oxford: Philip Allan.

O'Connell, T. and Kennedy, N. (1994) 'Dublin's International Financial Services Centre: a Review.' In D.E. Fair and R. Raymond (eds) *The Competitiveness of Financial Institutions and Centres in Europe*. Dordrecht: Kluwer Academic Publishers.

O'hUiginn, P. (1990) 'The International Financial Services Centre.' *The Irish Banking Review*, spring: pp. 31–33.

Raikes, D. and Newton, A. (1994) 'Competition and Financial Centres in Europe: London as a Case Study.' In D.E. Fair and R. Raymond (eds) *The Competitiveness of Financial Institutions and Centres in Europe*. Dordrecht: Kluwer Academic Publishers.

Revell, J. (1994) 'International Financial Centres in Western Europe.' Institute of European Finance Research Papers in Banking and Finance, no. 94/3, Bangor.

2

Financial Deregulation and Offshore Banking: Lessons for Malta from Australasian/Asia-Pacific Experience

Bruce Felmingham and James W. Dean

The Asia-Pacific region is among the fastest-growing in a global context. It is a region in which financial reforms have proceeded at a great pace as individual nations in the region perceive the benefits of financial integration. An important consequence of this financial reform is the competition among the island states of the region to establish offshore financial centre (OFC) status.

This striving for OFC status parallels in some respects the Maltese experience, which is developing the financial infrastructure required for this purpose. Our objective is to bring Australasian/Asia-Pacific experience of financial deregulation and offshore status to the debate about Malta's attempts to achieve full OFC status in a European context and to demonstrate how deregulation and offshore status are related.

We have structured our arguments in the following way. First, we describe economic trends in the Pacific Rim and in particular the economic integration of the region. Then, we analyse the financial reform agendas of nations comprising the region as a preliminary to a discussion of the development of offshore banking facilities around the Pacific Rim. Of particular interest is the emergence of Singapore and Hong Kong as major OFCs and the recent attempts by Thailand and Malaysia to compete in the offshore banking (OB) market. We analyse the economic benefits of deregulation in the next section. Then we examine the issue of offshore banking in the Asia-Pacific basin and in Australia and New Zealand. The lessons for Malta are drawn together and summarized in a closing section. They are presented in the form of costs and benefits flowing from the achievement of full OFC status. In general, we find that benefits exceed costs, although the costs warrant some discussion.

Recent Economic Trends in the Asia-Pacific Region

The East Asian nations

Drysdale (1988) summarizes the outstanding features of economic development in the East Asian region. The key to everything else is the rapid growth of the East Asian economies. Japan is the leading nation in the region: its gross national product at the beginning of the 1960s was less than 10 per cent of US GNP, but by the end of the 1980s the proportion had risen to 40 per cent. It had taken 30 years for Japan to achieve what the United States achieved in 90 years. Today, Japan is the world's second largest economy, third largest trader and largest capital exporter.

It is a mistake to think of rapid growth in terms of Japanese experience alone. The 'Asian tigers', namely Singapore, Taiwan, Hong Kong, Malaysia and Thailand, have average growth rates well above the Japanese rate at the same stage of development. In 1970, these countries contributed just 2 per cent of the world's GDP; by 1990 this percentage contribution had quadrupled.

The final piece in this jigsaw of rapid development is the comparatively recent emergence of China. The implications are discussed by Garnaut (1988). The People's Republic is the sixth economy in the north Western Pacific region to have doubled output within a single decade. None of the five preceding China failed to double output in the following decade. This

Figure 2.1   Australia, New Zealand and the Asian 'tigers'

experience has encouraged China's central planners to set ambitious growth targets for the year 2000.[1]

The amazing growth performance of these Asia pacific nations cannot be accidental; there must be some common explanations. Garnaut (1994, pp. 80–81) provides a succinct summary of these. The first common cause is the advantageous nature of the *geographical location* of nations in the region. All the six nations mentioned above are in comparatively close proximity and their development has been stimulated by the expansion of interregional trade. Close proximity means lower transport costs and a competitive advantage over nations exporting from a greater distance.

A second explanation of high Asia-Pacific growth rates is the *shared cultural tradition* of the region, often oversimplified to the description 'Confucianism'. However, there are elements of the Great Tradition of East Asia which relate to the rate of economic growth. these elements include the high value placed by East Asian communities on disciplined formal education, social cohesion and the capacity to accept unpopular reforms; the *work ethic of these communities*; and the common attachment to a growth ideology.

The form and style of government is one factor which does not seem to influence the pattern of growth. China is dependent on a Chinese interpretation of Marx and Lenin with a substantial legacy of central planning, while Japan is a parliamentary democracy which for years was governed by the same party, albeit at the expense of some political corruption in recent years. The foundations of Singapore's expansion were laid under a form of guided democracy, while Hong Kong has prospered as the world's model *laissez-faire* state. There is no common pattern to the governance of these high-growth economies.

Finally, the *size* of the East Asian economies does not impede growth. Singapore and Hong Kong are small city-states whereas China and Japan are Asian giants.

Hughes (1988) draws a stark contrast between the performance of the Australian economy and those in East Asia. At the turn of the twentieth century, Australia's per capita income ranked first among the world's economies; it was ranked third highest at the end of the Second World War, but had fallen to 18th at the end of 1994.

Hughes's explanation is consistent with an economic rationalist view: Australia persisted for too long with a fixed exchange rate, heavily protected markets and an excessive amount of regulation. Unlike East Asia, the form and structure of Australian government does matter for the growth rate. Australia, a nation of 18 million inhabitants, has three layers of government: federal, state and local. The consequence is a large amount of economic regulation – the non-uniformity of state and federal legislation in particular and the duplication of regulations across the three tiers of government.

Balanced against this was the immense natural resource base of the Australian economy. These natural resources provided a comparative advantage for Australia in exports of agricultural commodities in the first half of the twentieth century and in exports of coal, iron ore, copper, tin and other minerals in the second half. Porter (1990) is critical of nations such as Australia which depend excessively on a natural-resource-based comparative advantage. His point is that Australia failed to diversify its export base early enough and allowed its terms of trade to wind down in the face of tough competition for agricultural and mineral exports.

The population of New Zealand, a nation of 3.38 million, is a mixture of essentially British Europeans and Polynesians. It is distinct from Australia in preserving a British heritage in its migration programme. British influences were evident in all aspects of New Zealand society up to the 1980s.

New Zealand's economic malaise throughout the 1970s is summarized by Spencer and Carey (1988), who indicate that New Zealand's per capita income ranking has tumbled from seventh among the OECD countries in 1960 to the bottom of the table in 1984. The decade to the end of 1984 was not one for New Zealanders to recall with fondness. There was no economic growth in this period, current account deficits were excessive and inflation was well above OECD averages. Apparently, New Zealand's economic policies were completely inappropriate in these circumstances:

> Contributing to these problems was a range of government economic policies that distorted relative prices in New Zealand so that individual households and firms could not make economic decisions based on the full costs and benefits to the nation. . . . The government's financial policies also distorted relative prices, in particular by artificially discouraging net savings in New Zealand and by encouraging investment in some activities at the expense of others. (Spencer and Carey, 1988: p. 2)

The case for financial deregulation in both Australia and New Zealand was almost undeniable at the beginning of the 1980s. However, these two Australasian nations were in a completely different position in comparison with their Asian neighbours. The rapidly developing Asian countries looked to further deregulation as a way of providing more stimulus to development. The Australasian countries, however, looked to financial deregulation in a different context, namely to arrest an economic decline. The differing approaches to financial deregulation are outlined in the following section.

Financial Deregulation in the Pacific

Financial reform among the East Asian nations has focused on removing impediments to the free flow of international capital. This has come about as a result of the desires of at least some of them to play a major role in the international capital market.

The international evidence suggests that the Asia-Pacific nations are already large players in the international capital market; see Dean (1995), for example. The developing countries' share of global direct foreign investment has risen from 10 per cent in the 1980s to 40 per cent in the 1990s. The East Asian nations accounted for 50 per cent of these private capital flows.

This trend towards investment in the Asia-Pacific region is to be found in the various forms of indirect investment. The commercial banks have favoured this region in the writing of new foreign loan business; the emergence of the Asian dragon bond market accompanies a sharp rise in bond issues from the Asia-Pacific region.

Singapore, Hong Kong and Indonesia were well placed in regard to the mobility of international capital as early as the 1970s according to Garnaut (1994, p. 84). These countries operated fixed exchange rate regimes without any exchange controls. It is customary to associate exchange controls with fixed regimes and financial deregulation with the simultaneous removal of both fixity and exchange controls. Exchange controls, which take the form of restrictions on the convertibility of the domestic into foreign currency, are supposed to buttress the effectiveness of policy-inspired devaluations or revaluations of a fixed exchange rate.

It is testimony to the effectiveness of policy-making in these three countries that they were able to maintain fixed exchange rate controls. However, other countries in the region did impose exchange controls with mixed success. Thailand, for example, was able to maintain a different interest rate structure for a brief period with the help of exchange controls, but these controls proved ineffective over the medium term. Warr and Bhanupongse (1994) analyse the Thailand experience of exchange controls.

It is noteworthy that none of the East Asian countries saw the floating of their currencies as important aspects of financial reform. The following quotation summarizes the situation:

> none of the East Asian currencies opted to float their currencies on a continuing basis. Periods of floating sometimes followed crisis in external payments and were used by the authorities to guide the selection of a new parity: Indonesia in the major adjustment and rehabilitation of the mid 1960s; the Philippines from time to time; Hong Kong for a lengthy period culminating in the seizure of the US dollar as an anchor for stability in 1984 and China and Vietnam today. All chose to return to the perceived certainty of some version of a fixed exchange rate. (Garnaut, 1994, p. 84)

There are several explanations. First, several of the high-growth Asian tiger economies were running current account surpluses, so that there was no urgent need to adopt corrective measures such as a freely floating currency. Second, the major concern of these Asian countries about the exchange rate was the role it might play in economic transition; that is, the transition from a primary producing exporting country to one with a trade focus dominated by manufacturing exports. The concern here is with the real exchange rate as opposed to the nominal rate.

Dean (1995) indicates that the Asia-Pacific countries have preferred to fix exchange rates in a band allowing current account surpluses to accumulate as foreign reserves. Any potentially inflationary consequences are sterilized through open market operations.

The real exchange rate is subject to a variety of influences: domestic costs (in particular wages), employment policies, subsidies and the general productivity of the domestic economy, which in turn influence relative prices. To these we can add the international influences, including the nominal exchange rate. Given the difficulty of isolating the effects of a floating regime on the real exchange rate, the East Asian nations have preferred the greater stability afforded by fixed exchange rate mechanisms.

The other notable feature of East Asian development is the growth and diversification of the capital markets in some East Asian nations. Singapore, Hong Kong and Japan are major financial centres on the Pacific Rim. The impact of financial deregulation on the Japanese banking industry is starting to echo through the world's capital markets.

In summary, East Asian financial deregulation has focused on the liberalization of capital flows in preference to trade liberalization achieved through floating of domestic currencies. The outcome has seen the establishment of East Asia as one of the world's three great financial centres. We shall see presently what financial deregulation means for the offshore banking business in the Pacific Rim.

Recent Japanese experience

Contrary to this general Asian trend, Japan remained a heavily regulated economy throughout the 1980s and first half of the 1990s. The tensions caused by over-regulation were heightened by the economic problems confronting Japan in the 1980s and 1990s. When describing these tensions, the Japanese talk of the 'bubble' economy and the aftermath of the bubble's bursting. The bubble describes the rapid escalation of asset prices which dominated the five years 1983–87 and the bursting of the bubble, which saw prices crash. In Western economies, the bubble is synonymous with the bull run on equity markets in the same period. However, Japan also experienced a rapid escalation of property prices and land values which made the cost of doing business in Japan excessive.

The consequences of the bubble's bursting are outlined by the governor of the Bank of Japan in its *Quarterly Bulletin* (Bank of Japan, 1995). In essence, the collapse of asset prices exposed the balance sheets of some major financial institutions, leaving them with a number of very large non-performing loans. Another feature of the Japanese bubble, not so evident in other countries, was the rapid appreciation of the currency, which had the effect of driving manufacturing offshore. The Japanese refer to this as hollowing-out of the manufacturing base.[2]

Most domestic analysis of the Japanese economy acknowledge that Japan's recovery has been hampered by the need for further deregulation, in particular of the Japanese capital market. One Japanese analysis puts the issue in the following terms: 'Reappraising regulations and customary business practices ... would likely prove an effective means of preventing a hollowing-out of the Tokyo capital and financial markets' (Hondo, 1994, p. 25).

Financial deregulation has been forced on the Japanese by the growing significance of the yen in international financial markets. Hondo suggests several reforms. To encourage yen-denominated trade financing by Japan's exchange banks and currency dealers, he suggests the removal of eligibility requirements and tax revenue stamp impediments on yen-denominated banker's acceptance trade bills (yen-BAs). He explains the need to deregulate the short-term yen bond market in particular, in order to improve liquid convertibility and to expand the market among domestic residents. The long-term Euro-yen market should be liberated from remaining regulatory impediments. Hondo also perceives a need to link yen loans to yen-denominated imports and in this way to assist yen loan recipients to repay their debt, while protecting borrowers against currency risk. These are the key deregulatory measures required to restore the true value of the yen on international markets and to arrest the outflow of financial capital market infrastructure from Japan to Singapore and Hong Kong.

The lesson to be learned from Japanese experience is that nations that are slow to deregulate their capital markets suffer a capital outflow to countries that are on the cutting edge of the reform process. Although exchange controls were lifted in the mid-1980s, other necessary reforms were not completed.

The slow pace of financial reform, the high costs of doing business in Japan and the associated overvaluation of the yen encouraged large capital outflows, some proportion of which did not reflect the true market position of the Japanese economy.

The Japanese reaction to global pressures for reform led to a long list of reforms passed by the Japanese Diet in November 1995. These and the collective intervention of the G7 central banks in the currency market forced a 20 per cent depreciation of the yen against the US dollar.

Further, the Diet passed legislation to subvent the bad debts of the banks, which repaired most of the financial damage flowing from the corruption that impacted on the Japanese capital markets in the late 1980s and early 1990s.

Australian reforms

The Australian approach to financial deregulation is outlined by Felmingham and Coleman (1994, ch. 20). Australia's capital market moved from being one of the world's freest in the mid-nineteenth century to one of the most heavily regulated by 1940. The years 1965–81 included some tinkering with institutional reforms, but in general the major institutions were unaffected. Monetary policy operated on the basis of a series of secure asset/deposit ratios and non-market-oriented quantitative restrictions. The banking system was organized on the oligopolistic, branch banking basis and was highly concentrated: 75 per cent of Australia's banking business was controlled by seven major banking companies. Interest rate ceilings were imposed by the Reserve Bank (Australia's central bank).

The watershed in Australia's financial reform was the Committee of Inquiry, chaired by Campbell, into the Australian financial system (Australian Financial System Inquiry, 1981), which recommended widespread deregulation. Most of these reforms were implemented in the first half of the 1980s. Interest rate ceilings were removed; monetary policy assumed a market stance with quantitative approaches abandoned; the distinction between trading and savings banks was eliminated; and the favoured taxation treatment of the non-bank financial intermediaries was reduced. In other words, a competitive, level playing-field was created.

Australia's foreign exchange market was deregulated in December 1983 and all exchange controls abolished. The floating of the currency was inevitable given the inadequacy of the crawling-peg regime in accommodating large capital inflows and outflows and by the need for these to finance persistent current account deficits. In contrast to its East Asian neighbours, Australia was a current account deficit nation and the floating of the currency was viewed as a potential solution to the nation's deficit dilemma.

New Zealand's reforms: 'Rogernomics'

We have described the heavily regulated and subsidized status of the New Zealand economy in early paragraphs. However, the election of the Lange Labour government in 1984 brought to centre stage Roger Douglas, who embarked on a reform process which attracted worldwide attention under the label 'Rogernomics'. Financial deregulation was one of three reform agendas; the others concerned the privatization of state-owned enterprises and the removal of subsidies from these and the reform of labour market arrangements. Rogernomics shattered the tranquillity of life in this southern Pacific nation. New Zealand was dragged involuntarily into the last decade of the twentieth century as one of the most deregulated economies in a global context. Here we emphasize the financial reforms, which are covered in more detail by Spencer and Carey (1988).

The first measure taken by the new Labour government was to abolish controls on interest rates, allowing these to be determined by market forces. The second was logical: the floating of the New Zealand dollar and the complete abolition of exchange controls. New Zealand did not have any choice about this, having spent $NZ746 million defending an overvalued exchange rate in June and July 1984. New Zealand's approach to the revision of monetary policy was more radical than Australia's, and in some respects needed to be. New Zealand's peculiar brand of ratio controls, imposed by its central bank, had distorted the flow of savings into different financial institutions. These were abolished in February 1985. New Zealand lifted restrictions on bank registration to allow an unlimited number of players into the country's small bank market. Australia had issued 15 licences to foreign banks in 1985. New Zealand's Reserve Bank's prudential supervision of the commercial banks was formalized and regulations surrounding the operations of the nation's building societies and banking industry were revised to place these institutions right on the market.

A notable difference between the Australian and the New Zealand reforms was the treatment of the central banks of these countries, the Reserve Bank of Australia (RBA) and the Reserve Bank of New Zealand (RBNZ). The New Zealand reforms included a revision of the RBNZ charter to focus its monetary policy actions on a single objective, namely inflation control. The RBA's charter survived deregulation and does not refer directly to the control of inflation. Monetary policy targets include general economic welfare, full employment and the stability of the Australian dollar on foreign bourses. The debate in Australia about the RBA charter continues.

It seems certain that under a new coalition government (elected March 1996), Australia will remodel central bank operations along New Zealand lines.

What Benefits does Financial Deregulation Provide?

Has financial deregulation increased the economic well-being of the inhabitants of the region? We answer this question by examining the effects of financial reform on a key issue: the competitiveness of the nations on the Pacific Rim. We then cite evidence about the general macro-economic consequences before considering a further basic point: the effects on income distribution.

Effects on international competitiveness

One of the presumed consequences of deregulation is that it makes previously regulated economies more efficient and improves their capacity to compete on world markets. In time, increased competitiveness improves economic welfare through accelerated growth. Table 2.1 provides a comparison of indices of competitiveness in the Asia-Pacific region at three points in

Table 2.1   Changes in domestic price level expressed as an average of changes in US, German and Japanese price levels

Nation	Indices			Percentage changes	
	1980	*1987*	*1992*	*1980–87*	*1987–92*
Australia	74	100	81	26	-19
New Zealand	105	100	108	-5	8
Indonesia	50	100	99	50	-1
Malaysia	78	100	100	22	0
Philippines	69	100	83	31	-17
Singapore	84	100	79	16	-21
Thailand	75	100	90	25	-10
China	41	100	91	59	-9
Taiwan	98	100	75	2	-25
Hong Kong	76	100	64	24	-36
Japan	128	100	91	-28	-9
South Korea	78	100	78	22	-22
Canada	97	100	72	3	-28
USA	98	100	94	2	-6

Source: Garnaut (1994, p. 90). Reproduced with kind permission from the *Economic Record*

time – 1980, 1987 and 1992 – and the percentage change of these indices of competitiveness between 1980 and 1987, and 1987 and 1992. The index of competitiveness is an estimate of changes in the real exchange rate.

These changes in international competitiveness over the period 1980–87 simply reflect different stages of development and the effect of development on the real exchange rate. New Zealand's nose-dived in 1987 at the height of the structural reform process, while Japan's competitiveness, which had escalated steadily until 1986, fell sharply with real depreciation in that year in the face of burgeoning current account surpluses and world intervention. The outcomes for China (+69) and Indonesia (+50) reflect the rapid transition of these countries from being producers of primary exports to being producers of manufactures. The period 1980–87 is associated with the first signs of rapid growth in China.

By the mid-1980s, growth patterns were established, and in the cases of Japan, Hong Kong and Singapore the growth rate had stabilized following rapid expansion in the 1970s. Further, the financial reform agendas of most Asia-Pacific countries had been largely implemented, and if we were to witness the benefits then they should have been apparent in the period 1987–92. Real depreciation is evident in all countries over this period, owing to the influence of the recession, among other factors, but note how the competitiveness of Hong Kong, Canada, Taiwan, Korea and Australia fell more rapidly in comparison with other countries.

The outstanding characteristic of Table 2.1 is the performance of New Zealand, which defied the trend of real depreciation and increased its international competitiveness (by 8 per cent). Perhaps the radical reform agenda labelled 'Rogernomics' had worked and explains the relative improvement of New Zealand's international competitiveness.

General macro-economic consequences

New Zealand's financial reform agenda has contributed to an improvement in the country's competitiveness, but at what cost?

The answer is summed up in one phrase: income distribution.

Saunders (1993) concluded that the pattern of income distribution in Australia showed only small variations between 1942 and 1943 and 1989 and 1990. But later data for 1993–94 show a sudden shift in income distribution towards the higher-income group. This coincides with the substantial restructuring of the Australian economy and leads one commentator to lament this attack on a cherished Australian dream:

> Australians think of themselves as notably egalitarian, but the distribution of income before tax has the same characteristics in Australia as in Hong Kong: the lowest 20 percent of the population earns 5 percent of National Income and the highest 20 percent earn 47 percent. (Hughes, 1988, p. 188)

The New Zealand experience is no different from Australia's. There has been a discernible shift in the pattern of New Zealand's income distribution in favour of the higher-income groups. This is more easily comprehended in the New Zealand case. New Zealand had preserved a fairly even standard of living throughout the 1970s and early 1980s by maintaining zero unemployment over this period. In other words, everyone had a job but their real incomes were lower because consumer prices were inflated by a full-employment policy based on import controls. It was inevitable that labour, financial and foreign exchange reform would widen the gap between rich and poor, certainly in the restructuring phase.

It seems inevitable that prevailing patterns of income distribution will be disturbed by economic liberalization, and this is true in general of nations on the Pacific Rim.

Experiences of Offshore Banking on the Pacific Rim

One of the benefits flowing from financial deregulation is the prospect of expanding domestic banking activities in the direction of offshore banking (OB) facilities. OB activity is expanding on the Pacific Rim at a rate commensurate with the economic development of the region. We summarize these OB developments in the following paragraphs.

The outstanding characteristic of offshore financial centre (OFC) activity is the success of some of the smaller countries in establishing OFC status. The small city-states of Singapore and Hong Kong are two important examples. Although the uncertainty surrounding the future of Hong Kong as a trade and financial centre beyond 1997 has encouraged other small Asian nations to bid for a share of financial centre business, New Zealand's reform package has opened the way for a substantial increase in that country's share of the OB business. Finally, Australia has facilitated its OB business by legislating tax advantages. We focus on the OB business of these small Pacific nations because they carry the more significant implications for small island economies such as Malta or Cyprus.

Singapore is a shining light in the region, having expanded its capital market from purely domestic orientations towards full OFC status. The Singapore experience warrants special emphasis.

The first explanation of the Singapore story concerns the policies of the Singapore government, which in relation to the capital market have two major objectives: first, to desegregate the domestic and international aspects of the capital market, and second, to broaden the economic base while preserving the non-economic characteristics of Singapore. The international component of Singapore's financial markets was established through the provision of fiscal, regulatory and tax incentives to international financial institutions of established reputation. This proved successful, and several international banks established Pacific basin regional headquarters in Singapore.

The singapore government created two distinct banking units in 1978: Asian currency units (ACUs) and domestic banking units (DBUs). The ACUs were confined at first to offshore transactions while the DBUs dealt exclusively with domestic bank business. This created a formal distinction between international and domestic banking business. ACUs were subject to fewer regulations, lower taxes and less restrictive policy prescriptions. Tax concessions in particular attracted offshore business from other Asian countries. The distinction between offshore and domestic banking business was entrenched more deeply in 1971, when the government established restricted foreign banking licences. Foreign banks were confined to a single branch and were restricted in their capacity to accept Singapore currency deposits.

There are other explanations of Singapore's success in establishing offshore facilities. Singapore is strategically located in a time zone advantageous to the conduct of international financial transactions; it possesses an adequate supply of worker skills from its expanding educational base; and it has an advanced information technology infrastructure, essential for international banking.

The Singapore International Monetary Exchange (SIMEX) was established in 1984 and from this point Singapore was able to expand its international network of capital market links. SIMEX, for example, established mutual offset facilities with the Chicago Mercantile Exchange.

Hong Kong is another major OFC on the Pacific Rim. It makes for an interesting comparison with Singapore. Hong Kong is the closest of the global economies to the notion of a *laissez-faire* economy; Singapore's economy is closely controlled. Hong Kong as a British protectorate until 1997 does not offer foreign institutions the security of supervision by a central bank. Regulation is exercised by the Commissioner of Banking, and the Hong Kong and Standard Chartered banks issue notes and manage clearing-house operations.

The absence of a central bank distinguishes Hong Kong from Singapore, but this limitation has not impeded the development of Hong Kong as a financial centre on the Pacific Rim. In January 1993 Hong Kong was host to 527 banking firms. The 163 *licensed* banks operate the full range of commercial banking services, but the 53 *restricted licence* banks are referred to as merchant banks with minimum deposit requirements of HK$500,000, while 159 *deposit-taking companies* operate as mortgage finance and leasing companies. Hong Kong's status as an OFC is reflected in the fourfold ratio of foreign currency to domestic assets, while more than half of total deposits are in foreign currencies.

Hong Kong's future as an OFC remains in doubt until the intentions of the People's Republic are revealed. This uncertainty has slowed the growth of OB activities in Hong Kong and created a market opportunity for other potential players.

In 1993, the Thai government introduced offshore banking units (OBUs) in Thailand and gave them tax-exempt profits status on foreign-currency loans and reduced taxes on interest earnings to 10 per cent per annum. This has promoted Thailand's OFC status substantially.

Australia's banking industry, encouraged by the flexible exchange rate and removal of exchange rate controls, has diversified into OB activities. This was stimulated further by the decision in 1985 to license 15 foreign banks in the Australian market. Australia's banks found that foreign borrowing represented a cheap source of funding because it was not subject to statutory reserve deposit (SRD) requirements. This advantage disappeared with the abolition of the SRD system in 1988. This year is marked by a regulatory change to the operations of Australia's banking units, which were allowed to borrow and lend to non-residents in foreign currencies. A further attempt to stimulate Australia's OB business came with the announcement of further tax concessions for OBUs on 1 July 1993. This allowed concessional tax rates of 10 per cent per annum on OBU profits in place of the normal tax rate of 39 per cent. At the same time the state of New South Wales abolished stamp duty and financial institutions duty on OBU documents.

Australia's OB business has not prospered, partly because of this duplicate tax system. Income taxation is levied by the federal government, but the six state and two territory governments impose a confused mix of indirect taxes. A uniform approach to tax incentives for OBUs is not easily achieved in Australia. Presently, only 1 per cent of Australia's total foreign exchange business is conducted by OBUs.

One of the difficulties for Australia in developing the volume of its offshore banking business is the expansion of New Zealand interest in OB. The capital markets of Australia and New Zealand are integrated closely: Australian banks control more than half the New Zealand market and Australian banks conduct a significant and expanding proportion of their OB business through New Zealand branches.

The preference for the New Zealand option is explained by the higher returns earned by Australian banks on OB business conducted through New Zealand branches and agencies. New Zealand's financial liberalization agenda has proceeded at a faster rate and is not hampered to the same extent by fiercely independent regional governments. Australia's three-tier system of government continues to be a serious impediment to Australia's attempts at expanding the OB business.

Implications for Malta

In closing, we bring these experiences of financial deregulation and offshore transactions occurring in the Asia-Pacific region to developments in Malta. We do this by first considering the economic and social impacts of financial deregulation. We then turn attention to the issue of OB and OFC status; the costs and benefits of OB business are considered and a schedule of requirements for the successful establishment of a fully fledged international financial centre is listed.

Economic effects of deregulation

An important conclusion from our study of the Australasian and Asia-Pacific countries is that financial deregulation is an important precursor to the establishment of international OFC status. However, one aspect of deregulation stands out: the need to liberalize the flow of international capital. The achievement of OFC status depends centrally on the unfettered mobility of capital. All those nations on the Pacific Rim achieving OFC status have delivered this to the major international financial institutions. In addition, the host centres have offered taxation concessions as an incentive. So the free mobility of capital and taxation incentives is the common item on the financial deregulation agenda in the Asia-Pacific region, and it is this that has led to the stimulation of offshore financial activities.

There are, however, diverse approaches to financial deregulation in the Asia-Pacific region apart from these two common factors. Hong Kong and Singapore represent the leading example: Hong Kong is closer to *laissez-faire* capitalism than any other small open economy, while Singapore is a more closely and directly controlled economy. However, both have achieved OFC status and both have benefited in terms of growth and development.

Has financial deregulation increased growth rates in the Asia-Pacific region? The answer is a qualified yes. The general view of the New Zealand experience is that 'Rogernomics' has lifted New Zealand's economic performance. However, there is a dissenting view: 'The overall context is the massively awkward fact that three of the major macroeconomic indicators – growth, income distribution and especially unemployment – performed better over the quarter century before the reforms than they have done on average since 1984' (Hazeldine, 1993, p. 26). However, this argument is reversed if a more recent comparison is made. Growth rates, inflation and trend unemployment have all improved since 1982.

Nations such as Singapore, Thailand, Hong Kong, Korea and Malaysia, the 'Asian tigers', are among the fastest-growing nations in the world. And the same can be argued for Australia, which experienced the second fastest growth rate among the OECD group of countries in the 12 months to the end of October 1994. Not all this growth can be attributed to financial deregulation; however, the benefits to Singapore and Hong Kong of preserving a free market for international capital are obvious.

Effects on income distribution

Our analysis of the Asia-Pacific countries suggests that, in the short term at least, rapid financial reforms produce perverse outcomes for the pattern of income distribution. This is certainly true of the latest Australian and New Zealand data on income distribution in these countries. Financial reforms require the removal of subsidies to industry, the reduction of welfare payments and, in the case of New Zealand, the introduction of a regressive consumption tax. The outcome is summed up in the following: 'Many of New Zealand's tax and welfare reforms have proved controversial. Some, particularly the tax reforms, are seen to have clear efficiency advantages, but their overall impact on income distribution has been perverse' (St John, 1993, p. 41).

The message for Malta and for small island economies with narrow tax bases in general is clear: financial deregulation which improves efficiency may distort the pattern of income distribution in an undesirable fashion.

Non-economic considerations

The global economic imperative often poses threats for non-economic or cultural aspects of life. For example, in defence of the European Union's agricultural pricing policies, it is argued that the removal of farm subsidies would destroy the French farm and with it a vital aspect of European culture. So there must be some cautionary note entered about the need to preserve the uniqueness of the Maltese culture while the island's economy reaps the benefits of financial reform.

Singapore is a leading example of a small country which has achieved maximum economic benefits from financial reform and yet has striven to preserve an identity distinct from the excessive aspects of Western culture. It has been singularly successful in achieving both objectives and represents an important example for Malta.

The costs and benefits of offshore banking

We have assumed throughout that the development of OB business is beneficial, without putting this proposition to any test. Malta should not embrace the notion of becoming an OFC without some consideration of the costs and benefits involved. These are summarized by Valentine (1990) for Australia, but are worthy of consideration more generally.

Basically, the benefits take the form of the additional income which is earned from the establishment of an OFC. In the case of Singapore and Hong Kong, these income benefits are substantial. The costs take the form of a loss of government revenue involved in the provision of tax incentives and the prospects of tax avoidance. Clearly, financial planning should not promote undesirable activities. The reputation of some offshore centres has suffered as a consequence of these activities. The avoidance also reduces the prospects of benefits being spread widely. Valentine's conclusion is that the benefits exceed the costs in an Australian context, and they are likely to in the case of Malta as well.

From this analysis, we have compiled a list of those factors which we see as necessary for the establishment of a viable OFC. It is of interest to determine how many of these apply to Malta at present and which remain. This list of attributes is appended.

Appendix: the Requirements for a Successful Offshore Centre

Market structure	(Essential)	– An active offshore bank market – Taxation concessions – Mature foreign exchange market – Unrestricted capital mobility – Mature domestic banking system – Capacity for companies to establish regional HQs
	(Helpful)	– Developed financial future market – Equities market – Funds management facilities
Infrastructure		– Stable political environment

– Supervisory facilities (e.g. central banks)
– Strategic geographic advantages
– Attractive lifestyle and cultural diversity
– Legal, accounting and consulting services of sufficient quality
– Adequate local financial market skills

Question mark   – Exchange rate regime

Notes

1. These are described by Garnaut (1994, p. 82). The summary target is that GDP per capita in the People's Republic of China is targeted to quadruple by the turn of the century.
2. A reference to the movement of productive capacity offshore: labour-intensive, uncomplicated components being manufactured in less developed countries and capital-intensive components being concentrated in Japan.

References

Australian Financial System Inquiry (1981) (K.O. Campbell, chair), *Final Report*. Canberra: AGPS.

Bank of Japan (1995) 'Issues Facing the Japanese Economy and Roles of Central Banks.' *Quarterly Bulletin*, February: pp. 10–14.

Dean, J. (1996) 'Recent Capital Flows to Asia/Pacific Countries.' *Journal of the Asia Pacific Economy*, June.

Drysdale, P. (1988) 'Japan a Pacific and World Economic Power.' *Australian Economic Papers*, December: pp. 159–171.

Felmingham, B. and Coleman, W.O. (1994) *Money and Finance in the Australian Economy*. Sydney: Irwin.

Garnaut, R. (1994) 'The Floating Dollar and the Australian Structural Transition: The Asia Pacific Context.' *Economic Record*, vol. 70, no. 208: pp. 80–96.

Hazeldine, T. (1993) 'New Zealand Trade Patterns and Policy.' *Australian Economic Review*, vol. 40, pp. 23–27.

Hondo, K. (1994) 'Keeping the Yen down.' *Japan Scope*, winter: pp. 22–25.

Hughes, H. (1988) 'Too Little, Too Late: Australia's Future in the Pacific Economy.' *Australian Economic Papers*, December: pp. 187–197.

Porter, M. (1990) *The Competitive Advantage of Nations*. London: Macmillan.

St John, S. (1993) 'Tax and Welfare Reforms in New Zealand.' *Australian Economic Review*, vol. 4: pp. 37–42.

Saunders, P. (1993) 'Longer Run Changes in the Distribution of Income in Australia.' *Economic Record*, vol. 69, no. 207: pp. 353–366.

Spencer, G. and Carey, D. (1988) 'Financial Policy Reform: The New Zealand Experience.' Discussion Paper no. G88/1, Reserve Bank of New Zealand.

Valentine, T.J. (1990) 'Offshore Banking.' *Pacific Economic Papers*, vol. 121. Australia–
Japan Research Centre, Australian National University.
Warr, P.G. and Bhanupongse, N. (1994) 'Macroeconomic Policies Crises and Long
Term Growth in Thailand.' Washington, DC: World Bank.

3

Labuan, Malta and Belize: Evolution of Three Small Offshore Banking Centres

Eugene Sarver

A British Legacy Spanning Three Continents

London is not only the principal transaction centre of the Eurocurrency market, but its cultural heart as well. Reflecting London's dominance, the English language, British banking practices and English legal and accounting standards are entrenched throughout it, especially in its more classically offshore centres. Moreover, as 'financial services' generally represent a disproportionately large share of gross domestic product in such centres, the Anglo influence is heightened.

Such English-oriented offshore centres in Europe include Jersey, Guernsey (Channel Islands), Gibraltar, Cyprus, the Isle of Man and, of course, Malta. In Asia, such centres include Hong Kong, Singapore, Vanuatu (formerly the New Hebrides) and one subject of this study, namely Labuan, a federal district of Malaysia. Finally, in North America, such centres include Cayman, Nassau (Bahamas), Barbados, Antigua, the British Virgin Islands, Anguilla, the Turks and Caicos, Panama (via its US link) and, of course, Belize (previously British Honduras). This study will focus on offshore banking in Labuan, Malta and Belize – geographically and ethnically dispersed, but with an overlapping business culture.

The International Eurobanking Environment

The success of a particular offshore centre is influenced by the global level of cross-border financial activity, the growth rate of its own region, the relative attractiveness of large as compared with small centres, and its individual characteristics, including its political-economic stability, quality of life and special regulatory incentives. This section will address the three 'macro' factors initially listed, while the subsequent sections will discuss the individual centres of Labuan, Malta and Belize.

Statistics available from the Monetary and Economic Department of the Bank for International Settlements (BIS) in Basle, Switzerland (Bank for International Settlements, 1994) indicate a general slow-down in international banking as of mid-1994, with BIS reporting banks' gross international assets – the sum of cross-border claims and local foreign currency claims – to have declined by $78.7 billion, bringing the total cumulative fall since the end of 1993 to $89.6 billion. Notably, this trend excluded both London and the offshore centres in Asia and North America. The slow-down in Europe, which inevitably affects Malta, reflects the sluggish European economic recovery, with evidence of the soft market seen in an easing of lending terms, such as spreads, credit standards and maturities. Conversely, the Asian offshore centres have benefited from the strong growth of the Pacific Rim economies, while the North American offshore centres have likewise benefited from the strong economic growth of the United States and the recovery of Latin America.

In the wake of the 1991 Bank for Credit and Commerce International (BCCI) débâcle, involving the massive bankruptcy (in the end it had only $5 billion worth of assets against $20 billion in liabilities) of a bank which had branches in virtually every offshore centre (70 different countries), aside from the Channel Islands and Dublin (which wisely would not grant it branch licences to accept local deposits), there developed an aversion to the smaller Euro-centres. The latter reflected BCCI's abusive use of the Cayman Islands and Luxembourg for its nominal dual headquarters, and the growing awareness of the inability of small centres' authorities to monitor effectively the activities of sophisticated global banks, a consciousness enhanced by the new guidelines promulgated by the Committee on Banking Supervision (the 'Cooke Committee') of the Bank for International Settlements, implemented in the United States by the Federal Reserve Board on the basis of the Foreign Bank Enhanced Supervision Act 1991.

However, by 1994 there was an apparent recovery of confidence in smaller centres (much as confidence returned to the Euromarkets a couple of years after the Bankhaus Herrstatt, Cologne, and Franklin National Bank, New York, bankruptcies of 1974). Part of that renewed confidence reflects tighter regulation in the smaller centres to comply with the new BIS guidelines, as well as a more modest and careful attitude on the part of the smaller-centre governments to avoid involvement with international criminals, such as drug money launderers, embargo violators, financial manipulators, etc.

The Evolution of Labuan's Offshore Centre

Labuan is a five-year-old offshore banking centre located in a port city of 35,000 inhabitants (predominantly ethnic Chinese) on a 92-square kilometre tropical island, just 9 kilometres north of the east Malaysian state of Sabah, which serves as an entrepôt for the north Borneo region (see Figure 3.1). It is a federal territory directly administered by the government of Malaysia, and officially declared an international offshore financial centre (IOFC). Labuan has the unique advantage of providing privileged entry into the banking sector of the fast-growing Malaysian economy, the second strongest in the region after Singapore, with annual growth rates of over 8 per cent and a per capita GDP of over $2,800. Relatedly, Malaysia has low unemployment (3 per cent), favourable inflation (below 4 per cent) and merchandise trade (but not current account, with its $6 billion deficit), capital, and a balance of payments surplus, which in 1992 augmented its international reserves by $10 billion to a total of $27 billion (collectively, this resulted in high confidence in Malaysia's stability, reflected in its Moody's sovereign rating of A-2). Additionally, Labuan is very close to the oil-rich Kingdom of Brunei Darrusalam (whose ruler is the world's richest man), and is just 30 minutes by air from Sabah's capital of Kota Kinabalu, from where there are flights to all major Asian capitals. It is also within 1,500 kilometres of all the capitals of member states of the Association of South East Asian Nations (ASEAN) except Bangkok.

Amenities include luxury hotels, a major office–expatriate residence complex just completed (Financial Park), an airport undergoing a $75 million expansion, a telephone system being enhanced by a $28 million upgrade (including 12,000 new telephone, telex and data lines), and the local availability of the Reuters and Telerate telex and electronic information services. Offsetting factors include Labuan's somewhat isolated position and corre-

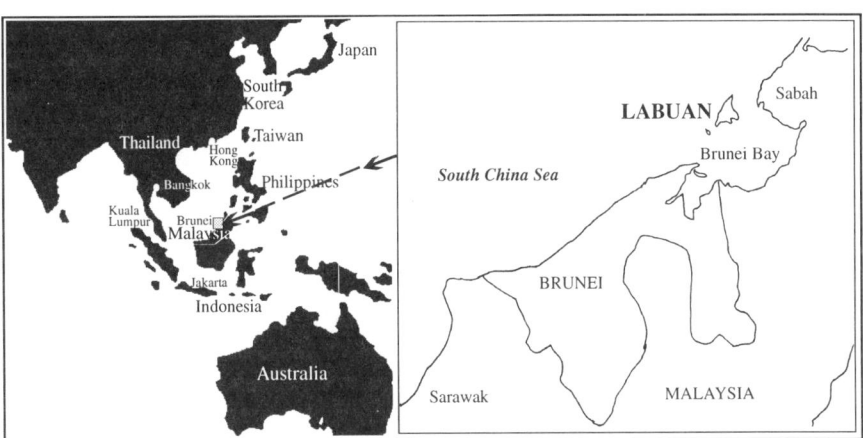

Figure 3.1 Labuan and its location near Brunei in the South China Sea

sponding transportation inconveniences, suboptimal physical infrastructure, subdued ethnic tension between the Malays and ethnic Chinese (moderated by the economy's strong growth, currently 9.2 per cent), and some political tension in Malaysia, including the secessionist sentiment in Sabah and short-lived media and international organization disputes with the United Kingdom and Australia.

Beyond providing an attractive physical infrastructure for standard off-shore financial activities, the Malaysian government has passed an extensive series of laws providing special incentives for priority activities. The legal infrastructure establishing Labuan as a tax haven was based on legislation including the Labuan Offshore Business Activity Tax Act 1990, the Offshore Companies Act 1990, the Labuan Trust Companies Act 1990, the Offshore Banking Act 1990, the Offshore Insurance Act 1990 and the Income Tax (Amendment) Act 1990. Pursuant to the latter legislation, aside from 40 banks, five insurance/reinsurance companies have been established (including two from Hong Kong and one from the United States), one (Malaysian) fund management company has been initiated, and 250 'other trading and non-trading offshore companies' have been established.

Incentives specifically for banks (and their depositors) include exemption from reserve requirements and foreign exchange controls, a low ($22,222) annual licensing fee (about half that of Singapore), and taxes of just 3 per cent of net profits, or a fixed rate of $7,400, whichever is elected by the firm. It is also more profitable to lend to Malaysia through Labuan because offshore banks there are exempted for the withholding tax levied by the government on income or dividends earned in Malaysia and remitted out. Moreover, 'non-trading' activities, such as the ownership of securities and real estate, are totally tax-exempt. Additionally, there are no stamp duties (for which Switzerland is notorious), value-added taxes (VAT), or death, inheritance or estate duties. Finally, income earned by an alien from employment in a managerial capacity in an offshore bank is 50 per cent tax-exempt until the end of 1997.

Responding to Labuan's incentives, three Malaysian banks – Maybank International, BBMB International Bank and Public Bank – became the initial Labuan licensees. In September 1992 they were joined by the Hong Kong and Shanghai Banking Corporation, which thereby became the first foreign licensee. As of 12 January 1995, as shown by Table 3.1, a total of 40 offshore banks had been established in Labuan, including 15 in 1993 and 19 more in 1994. In terms of country of origin, 13 banks are from Japan, 7 are from Malaysia itself, 4 more are from the United Kingdom and France respectively, 3 are from Singapore and the United States respectively, 2 are from Germany and the Netherlands respectively, and there is 1 each from Hong Kong and Switzerland.

In response to the incentives offered by the Malaysian government, as well as the proliferation of foreign banks in Labuan, there has been very strong

Table 3.1 Offshore banks in Labuan*

France
Banque Nationale de Paris, Labuan Branch
Banque Paribas, Labuan Branch
Banque Indosuez, Labuan Branch
Société Générale, Labuan Branch

Germany
Dresdner Bank AG, Labuan Branch
Bayerische Landesbank Girozentrale, Labuan Branch

Hong Kong
Dao Heng Bank Ltd, Labuan Branch

Japan
Mitsubishi Bank Ltd, Labuan Branch
Bank of Tokyo Ltd, Labuan Office
Fuji Bank Ltd, Labuan Branch
Tokai Bank Ltd, Labuan Branch
Sumitomo Bank Ltd, Labuan Branch
Long-Term Credit Bank of Japan Ltd, Labuan Branch
Industrial Bank of Japan Ltd, Labuan Branch
Dai-Ichi Kangyo Bank Ltd, Labuan Branch
Sakura Bank Ltd, Labuan Branch
Sanwa Bank Ltd, Labuan Branch
Daiwa Bank Ltd, Labuan Branch
Yasuda Trust and Banking Co. Ltd, Labuan Branch
Asahi Bank Ltd, Labuan Branch

Malaysia
BBMB International Bank (L) Ltd
Public Bank (L) Ltd
Maybank International (L) Ltd
D & C Bank (L) Ltd
UMBC International Bank (L) Ltd
Bank of Commerce (L) Ltd
AMMB International (L) Ltd

Netherlands
ABM AMRO Bank NV, Labuan Branch
Rabobank Nederland, Labuan Branch

Singapore
Overseas-Chinese Banking Corporation, Labuan Branch
The Development Bank of Singapore, Labuan Branch
United Overseas Bank Ltd, Labuan Branch

Switzerland
Union Bank of Switzerland, Labuan Branch

United Kingdom
The Hongkong and Shanghai Banking Corporation Ltd, Offshore Banking Unit, Labuan
Standard Chartered Offshore Labuan
Schroders Malaysia (L) Ltd
National Westminster Bank plc, Labuan Branch

United States
Citibank Malaysia (L) Ltd
Bank of America NT & SA, Labuan Branch
Bankers Trust International plc, Labuan Branch

Source: Bank Negara Malaysia (central bank) (1995)
* As of 12 January 1995

Table 3.2   The Labuan offshore banking industry

	31 Dec. 1991	31 Dec. 1992	31 Dec. 1993	30 Nov. 1994
Total deposits (US$ million)	317	1,005	7,730	2,480
Total loans (US$ million)	252	617	2,364	5,004
Total number of banks	5	8	21	40

Source: Bank of Negara Malaysia (central bank) (1995)

growth of the Labuan offshore banking industry as measured by total deposits and total loans. Specifically, as depicted in Table 3.2, in the approximately three-year period from 31 December 1991 to 30 November 1994, total deposits jumped 800 per cent from $317 million to $2.5 billion, while total loans soared 2,000 per cent from $252 million to $5.0 billion.

The Evolution of Malta's Offshore Centre

Malta is a 316-square kilometre island country (comprising the inhabited islands of Malta, Gozo and Comino, plus two uninhabited islands) with a population of 370,000, located 100 kilometres south of Sicily in the very centre of the Mediterranean Sea (see Figure 3.2). It was a British colony from 1802 until 21 September 1964, when it achieved full independence (after periods of limited self-government). Traditionally its well-managed economy, as reflected in 5 per cent growth, 2 per cent inflation, 4 per cent unemployment, current account equilibrium, a firm currency, international reserves of US$2.4 billion and a Moody's sovereign rating of A-2, has been based on tourism (1.1 million arrivals per year), light manufacturing (footwear, food processing, clothing, electronics), and shipbuilding and repair in its famous naval yards. However, it has now made a strong commitment to develop offshore banking and related activities by its passage of the Malta International Business Activities Act 1988. A large structural merchandise trade deficit of $600 million annually and the desire to absorb its growing labour force in higher-value professions were among the factors behind that commitment to develop Malta into 'the Switzerland of the Mediterranean'.

Some of Malta's general advantages for offshore banking, besides its central geographic position – equidistant between Europe (Italy) and Africa (Tunisia/Libya), and Gibraltar and the Middle East – include a stable political-economic environment (with a conservative Nationalist Party Prime Minister, Dr Edward Fenech-Adami, and a per capita GNP of $7,500); a large, highly qualified, motivated, multilingual (English, Italian, French) but low-cost labour force; schools granting highly respected diplomas in accounting, insurance and banking; and a high quality of life, including a sunny, moderate climate, available good-quality, low-cost housing, hotel and office space,

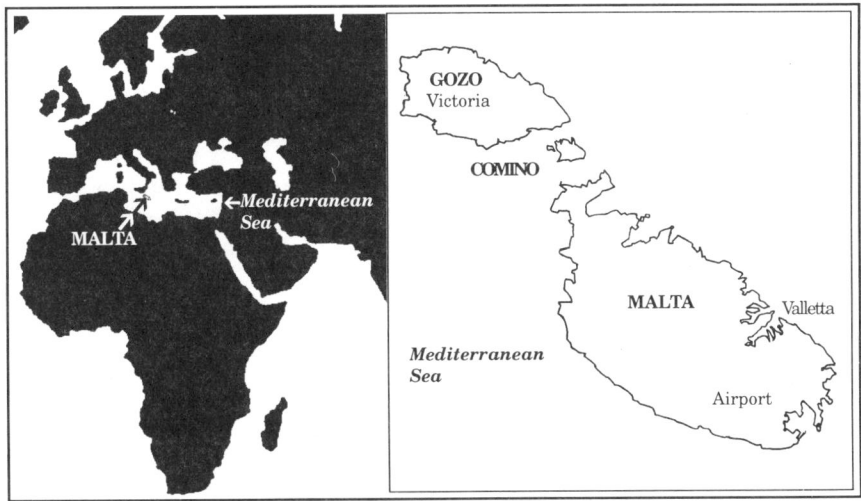

Figure 3.2 The Maltese Islands and their location in the Mediterranean Sea

a rich historical heritage/cultural environment (the Knights of St John commissioned Europe's finest architects in the late sixteenth century to build the capital, Valletta, and there are many Roman and even earlier sites), international cuisine, excellent schools for expatriates' children, recreational facilities, and a low crime rate.

Additionally, Malta is a traditional entrepôt as a consequence of its location, fine harbours, dry docks and warehouses, plus the favourably amended Merchant Shipping Act and complementary freeport operation (at Kalafrana). Moreover, Malta has something of a captive potential market from its annual 1.1 million tourists (from the United Kingdom, Germany, Italy, Libya, etc.), from its sizeable wealthy, largely English, expatriate community, and from the sizeable Maltese expatriate communities in the United Kingdom, Australia and the United States. Also, Malta has cultural ties to the Arab world, including its language, derived from ancient Phoenician, a Semitic language similar to Arabic; many of its towns, such as Rabat, have Arabic names despite its 97 per cent Roman Catholic population.

Finally, there is the mixed blessing of Malta's relationship to the European Union. Since 1970 it has had an association agreement and in 1990 it applied for full membership, with the European Commission in June 1993 confirming Malta's eligibility for eventual accession, which is expected in 1999. It is a mixed blessing because while full membership would give Malta full access to the 15-plus-nation European Union market (370 million-plus people, $7.3 trillion-plus GNP), it would subject Malta to the uniform financial regulations (e.g. restricting confidentiality related to income taxation) emanating from Brussels (European Financial Area or EFA), though with some possible exemptions reflecting its unique situation. Beyond the latter, Malta could use the 'Luxembourg strategy' of only providing bank account information on the basis of *judicial* as opposed to *administrative* (i.e. tax authorities) requests, which effectively protects the confidentiality of individual accounts except where a substantial conspiracy is involved. At any rate, Malta has made a full commitment to EU membership, and its Banking Act 1994 brought its banking laws into line with the EU standards; likewise, it modified its tax system to follow the EU's VAT system.

Specific incentives for offshore banks and their depositors include exemption from all foreign exchange controls (except in dealing with Malta residents) and exemption from regulations covering liquidity ratios, solvency margins, inadmissible assets and reserve funds as long as the Maltese entity is a subsidiary of a bank 'of international standing and repute'. Also, there is exemption from stamp duties, death duty and donation duty, and customs duties on the importation of an offshore bank's equipment and the personal belongings of its non-resident employees. Finally, there is just a 5 per cent rate of income tax, with the right of self-assessment, and a high degree of confidentiality.

Some disadvantages of Malta include its relatively late entry into offshore banking in the European region, where many other offshore centres have long been established; some political wariness from the disturbing policies of the prior Mintoff Socialist government (till December 1984); a scarcity of highly sophisticated international banking specialists; a perception of bureaucratic indifference (which the Malta Financial Services Centre or MFSC, recent successor to the Malta International Business Authority or MIBA, is trying to remedy by its exceptional accessibility) fostered, for example, by long, unnecessary delays at airport immigration; some wariness regarding MFSC's unconventional 'central bank role' relative to offshore banks; and the prohibition of onshore banking by offshore banks, denying them access to the large Maltese firms and the local expatriate community.

The small offshore sector is largely composed of the offshore units of Malta's two leading banks, the Bank of Valleta (Bank of Valletta International), Malta's largest stockholder-owned bank, and the $2.1 billion Mid-Med Bank (Mid-Med Bank (Overseas) Ltd), governmentally owned and the largest bank. For the Maltese banks, the offshore sector has become an important source of funding for domestic lending activities. Reflecting this, whereas during 1993 'foreign liabilities' rose 11.9 per cent to Lm 198.2 million (US$506.1 million), corresponding 'foreign assets' were virtually unchanged at Lm 110.9 million ($283.2 million). The difference, of course, is 'claims on the local commercial banks', which registered a 22.1 per cent increase to Lm 96.8 million ($247.2 million), as the two offshore subsidiaries of the major domestic banks continued to transfer funds emanating from 'non-resident deposits' to their parent banks in Malta.

Table 3.3 Malta: international banking institutions

	1994 Q4*	1995 Q1*
Deposit liabilities	Lm 104.8	Lm 162.1
Domestic assets	Lm 133.8	Lm 128.9
Foreign assets	**	Lm 61.3

Source: Central Bank of Malta, *Quarterly Review*, June 1995
* End of quarter.
** Not separately identifiable until 1995.

These Maltese banks are complemented by four foreign banks including the Melita Bank International Ltd (a subsidiary of Sanpalo Bank Holding Spa., Italy), the First Austrian Bank, the Belgian $20-million capital Izola Bank Ltd, and the First International Bank Ltd, a branch of the large, conservative Turkiye Garanti Bankasi AS (the first licensee for establishment in Malta of a bank not incorporated in Malta). The latter three are designated 'international banking institutions' (IBIs) and are allowed to transact business only with non-residents. At the end of June 1995, as shown in Table 3.3, the three IBIs had a consolidated balance sheet (domestic plus foreign assets) of Lm 190.4 million ($543.0 million).

To obtain a more detailed profile of financial flows in the Maltese offshore market, activity in the first half of 1994 was subjected to greater scrutiny. It revealed that the offshore banks' 'foreign liabilities' – partially reversing the 1993 gain – fell by 2.4 per cent to Lm 198.1 million ($501.4 million) by the end of March 1994. As in the final quarter of 1993, this decline was mainly due to a fall in 'balances due to other banks abroad' rather than to a fall in 'non-resident deposits', which, in fact, rose marginally. Meanwhile, the offshore banks' 'capital and reserves' increased by 3.8 per cent to Lm 4.5 million ($11.4 million), probably as the result of transfers to reserves. Concurrently, the offshore banks' 'claims on commercial banks' rose by 5.3 per cent to 96.8 million ($245.0 million), mainly as a result of the transfer of funds from 'non-resident deposits' placed with them to their parent banks. At the same time, the offshore banks' 'foreign assets' declined by 9.6 per cent to Lm 110.9 million ($291.8 million), reflecting the fall in their 'foreign liabilities' mentioned above.

Recovery in the deposit-taking business of the offshore banks characterized the second quarter of 1994, as the 'foreign liabilities' of the sector grew by 8.3 per cent to Lm 214.5 million ($576.1 million); it followed publication of proposed legislation concerning the setting up of the Malta Centre for Financial Services, as well as a related advertising campaign. Concurrently, reflecting the increase in their 'foreign liabilities', the offshore banks' 'claims on the commercial banks' rose by 12.8 per cent to Lm 105.7 million ($283.9 million) during the quarter, as the offshore banking subsidiaries of the Bank of Valletta and Mid-Med Bank continued to transfer funds obtained from non-resident deposits to their parent companies. The offshore banks, however, also increased their 'foreign assets' by 5.2 per cent to Lm 116.7 million ($313.4 million) during the quarter.

Developing Malta into a true international financial centre, of course, requires diversifying beyond banking into areas such as fund management. An important element of this strategy was implemented in May 1995 when Valletta Fund Management Ltd was initiated as a joint venture of the Bank of Valletta (57.5 per cent), British Rothschild Asset Management Ltd (37.5 per cent), APS Bank Ltd (2.5 per cent) and Lombard Bank (Malta) Ltd (2.5 per cent).

Finally, Malta's offshore banking sector is strengthened by a strategy of parallel development of offshore services beyond banking and fund management. For example, as of the end of 1994, a record 471 companies were registered under the Malta Financial Services Centre Act, half of which were trading companies, Malta's ship registry ranked tenth in the world (over 16.5 million tons), and its freeport is being expanded to handle a million containers a year.

The Evolution of Belize's Offshore Centre

Belize is a Central American nation of 22,965 square kilometres, bordered by Mexico and Guatemala, with a population of 200,000, 45,000 of whom live in the major port of Belize City (see Figure 3.3). A British colony until September 1981, it is now a parliamentary democracy headed by Prime Minister Manuel Esquivel of the United Democratic Party (UDP). Its economy has traditionally relied on exports of sugar, bananas and citrus products, predominantly to the United States (46 per cent) and the United Kingdom (31 per cent), as well as on tourism (250,000 arrivals per year) and light manufacturing (clothing and food processing).

Unfortunately, the country has significant economic vulnerabilities related to the withdrawal of its 1,500 British troops in 1994 (present as a result of a border dispute with Guatemala), which represented over 5 per cent of the gross domestic product, and to the susceptibility of its agricultural sector to adverse weather and trade barriers in the European Union and United States. Thus, it has developed a strategy of emphasizing offshore services, beginning in 1989 when it passed a law (Registration of Merchant Ships Act 1989) which created the International Merchant Marine Registry of Belize (IMMARBE), allowing merchant vessels from around the world to sail under its blue, red and white flag. Follow-up legislation in 1990 (the International Business Company Act) authorized international business companies (IBCs) – anonymous, tax-exempt entities that can be formed with just US$100 from outside Belize; already more than 1,800 IBCs have been set up. Finally, the Trust Act, enacted in May 1992 to provide for the creation of trusts, is recognized as one of the most modern of such legislations.

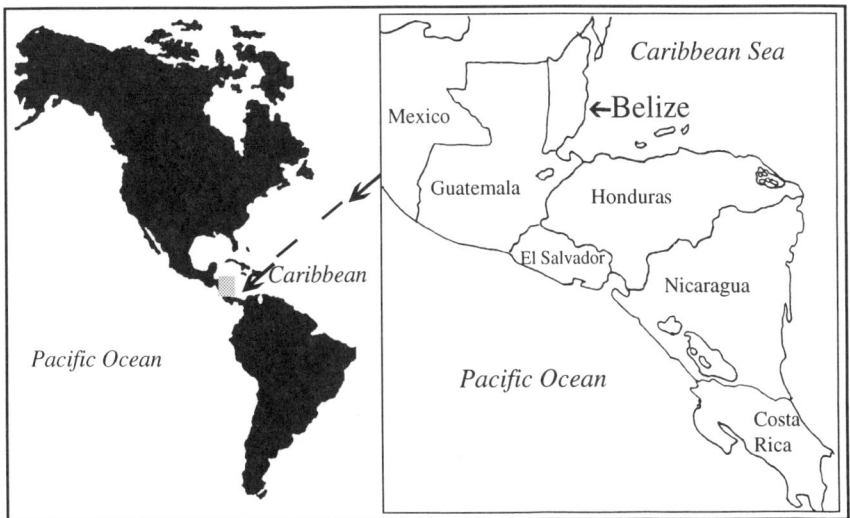

Figure 3.3 Location map of Central America and Belize

Although Belize has developed a wide range of traditional tax-haven activities, as indicated above, it has been slow and cautious with respect to offshore banking. This is because of a recent study sponsored by the British government which stated that the absence of a strong watchdog agency has left the nation vulnerable to criminal activity. In particular, it emphasized that the Central Bank lacks the expertise to monitor offshore finance, and that the few controls that do exist are in the hands of private business. As for the police, they are particularly unqualified to deal with sophisticated financial crimes. Moreover, the study found that the Colombian drug cartels have moved cocaine on Belizean-flagged vessels, and IBCs have been used to conceal the identity of businesses involved in breaking UN sanctions. Money-laundering seems also to be on the rise.

During 1996 the Belize National Assembly passed the Offshore Banking Act and the Money Laundering (Prevention) Act, both of which went into effect on 1 August 1996. They strike a judicious balance between the desire to attract foreign investment and the need to preserve the integrity and sound-ness of the financial system. As of May 1997 none of the many applications for offshore licences has been approved. While there has been a limited amount of such offshore activity, it had been on the awkward basis of *local* banking legislation. Specifically, as of the end of October 1994, Central Bank of Belize figures show that total 'non-resident foreign currency deposits' totalled only BZ$20.6 million (US$10.3 million), apparently almost all on deposit at Belize Bank, the leading bank. Concurrently, 'non-resident local currency deposits' (Belize dollars) totalled only BZ$26.4 million ($13.2 million).

However, current legislative activity indicates that the Belize Parliament will pass legislation in 1995 or 1996 to formalize offshore banking to enable Belize to offer the full range of standard offshore services, and thereby be more competitive with other Caribbean region centres such as Nassau, Cayman and Panama. In particular, the Belize Offshore Practitioners Association, led by its president, attorney Rodwell William, is aggressively lobbying for it to make Belize a full-service centre. While this will probably become a reality, the change will be made only in the context of a tightening up of regulation of the entire offshore sector and the requirement of frequent audits.

For offshore banking, Belize offers the advantages of a central Caribbean location next to the large Mexican economy; political and economic stability; a multilingual, educated workforce whose members often speak Spanish as well as English; and a satisfactory physical and legal infrastructure. Reflecting its British legacy, for example, the legal system is founded on English common law, and the court system is modelled on the English system, with the court of final appeal on questions of law being the Privy Council of the British House of Lords. Moreover, related to its strong tourism industry (including eco-tourism linked to its rain forests and its barrier reef, the world's second largest after Australia's), it has an excellent air service and telecommunications, hotel, restaurant and recreational facilities. Also, Belize has a relatively sophisticated private banking sector that includes domestic banks such as the Belize Bank Ltd (formerly the Royal Bank of Canada), complemented by British (Barclays Bank plc), Canadian (Bank of Nova Scotia) and Mexican (Banca Serfin) banks. Finally, Belize receives continuing economic and diplomatic support from the United Kingdom.

Negative factors include its relatively isolated location; a suboptimal physical infrastructure; some fiscal instability, with the government deficit an excessive 6 per cent of GDP (generally 3 per cent is considered the ceiling of responsible fiscal policy), though improvement may come after the introduction of a VAT system in April 1996; 'official international reserves' at an inadequate level of US$21.2 million on 31 March 1995 (a $6.9 million drop as compared with the end of the previous December); growing street crime; lack of offshore banking legislation combined with central bank supervisory inadequacies; and some border disputes with Guatemala (which theoretically were resolved until the Guatemalan government which signed the border agreements was overthrown).

Assessment of Commonalities, Disparities, Approaches and Performance of the Three Eurobanking Centres

As already shown, Labuan, Malta and Belize share many commonalities. All three are former British colonies, with their citizens using English as their first or second language. Beyond that, they generally adhere to British business practices, accounting and legal standards, as well as partaking of the more diffuse aspects of British business culture.

Second, all three are small and centrally located within their respective regions. That, combined with an education and sophistication level higher than that of many of their neighbours, has naturally drawn them towards entrepôt activity, such as import–export, light assembly, packaging, ware-housing, wholesaling and after-sales service activity. Inevitably, as the global economy has shifted away from manufacturing and basic industries towards services, these centres have expanded their service sector, and have increas-ingly been looking towards offshore services as an attractive and natural industry to piggyback on top of their existing range of international ser-vices.

Thus, all three began passing legislation (obviously the Malaysian national government did so in the case of Labuan) to create the legal infrastructure for an offshore tax haven centre, complemented by related improvements in the physical infrastructure, such as transportation and communication facilities. For all three centres, such activity began around the beginning of the 1990s, making them all comparably young centres, and therefore all relatively late entries to the already crowded global offshore banking market.

Being late entries, however, has given all three the advantage of being able to learn from the experience and performance of the existing centres in order to create an optimal legal, physical and business environment – although conversely it should be noted that being late has imposed upon them the burden of having to foster the early growth of their centres in the wake of the BCCI débâcle and related legislation linked to the resultant tough BIS guidelines, which, of course, adds to the challenge.

Not surprisingly, the studies by all three governments to develop an optimal strategy came to the identical conclusion that an offshore centre's viability and growth prospects are substantially derived from offering a wide variety of standard services that synergistically and symbiotically strengthen each other, and the underlying business sector of financial, legal and techni-cal specialists that supports them. Thus, in addition to offshore banking, they all generally offer such related services as international company registra-tion, trust companies, insurance, reinsurance and related brokering, and ship registration, among others.

While all three centres have ambitious plans and goals, they are all pursuing relatively *conservative* policies, wary of BCCI-style banks, narcotics traffickers and financial schemers. That reflects their conservative business cultures, partially derived from their comparably conservative and family-oriented cultures (Catholic in the case of Malta and Belize, and Muslim in the case of Malaysia, including Labuan).

While the commonalities among the three centres substantially over-shadow their differences, there are some significant differences as well. First,

Malta and Belize are tiny countries, while Labuan, the smallest centre, is paradoxically part of the country with by far the largest population (20 million). Second, Labuan (Malaysia) and Malta have relatively advanced economies compared to Belize, and therefore more governmental and financial resources to allocate to the development and marketing of offshore activities. Third, not only are they in very different regions, but more significantly those regions have significantly different growth performances and prospects. In particular, Labuan benefits from being in the strongest growth region of South-East Asia, while Belize benefits from the greatly improving prospects of Latin America (and strong US growth), subject to the fall-out from the recurrent Mexican currency crises, which led to diminished growth prospects in light of higher interest rates, fiscal austerity and reduced capital inflows. Concurrently, Malta is linked to the recovering, but slow-growing, European region and the problematic situation of North Africa, including Libya's isolated government and the military–Muslim fundamentalist strife in Algeria, which also affects Tunisia.

A related difference is the disparate competitive environment of the three centres. Malta, in the middle of the highly developed European region, faces the greatest competition. Classical offshore centres in the region include the Channel Islands, Luxembourg, Cyprus, the Isle of Man, Gibraltar, Madeira (Portugal), the Canary Islands, Monaco and Liechtenstein, and there is the beginning of activity in Morocco (Tangier) and Tunisia (Tunis). Concurrently, Malta must compete with the substantial Eurobanking sectors of the United Kingdom (London), Switzerland (Zurich, Geneva and Lugano), France (Paris), Spain (Madrid), the Netherlands (Amsterdam), Belgium (Brussels), Ireland (Dublin) and Denmark (Copenhagen), and the initiation of activity in Germany (Frankfurt). Additionally, Bahrain, Dubai (UAE) and Beirut are competitors for its Arab business. Thus, Malta is a new entry in an already saturated market.

Belize faces competition from its seasoned competitors in the Caribbean, especially Nassau and Cayman, but also Panama (diminished by its tarnished image), Barbados, Antigua, the British Virgin Islands, Anguilla and the Turks and Caicos. Also, there is competition from the Eurobanks, called international banking facilities (IBFs) in the United States (especially the Miami ones run by Spanish-speaking, Cuban-American bankers, and the New York ones), from the Eurobanking units of the giant Canadian banks in Toronto, Montreal and Vancouver, and from growing Eurobanking activity in Montevideo, Uruguay.

Finally, the Labuan centre faces competition not only from traditional offshore centres, especially Singapore and Hong Kong, but also from newer ones such as Taipei, Manila and Port Vila (Vanuatu), as well as Eurobanking units in Japan (Japan Offshore Market or JOM) and in Australia (Sydney and Melbourne). However, its regional competitors have significant weaknesses that improve its competitive position. For example, Singapore suffers from some notorious restrictions on information flows and has become relatively expensive and rigid, though with a high Moody's rating of AA-2. Concurrently, Hong Kong has a Moody's rating of only A-3 and reverts to the People's Republic of China on 1 July 1997, which is expected to clamp down on Hong Kong's free-wheeling culture and unrestricted information flows, and Taipei, while economically strong, as reflected in its Moody's rating of AA-2, is undermined by the 'two-China' question, and some political strife between its national 'mainland' government and the native Taiwanese. Finally, Manila, though improving, is weakened by the Philippines' frequent political instability and its balance of payments problems (which resulted in Wells Fargo Bank's famous lawsuit against Citibank, when its $2 million deposit at Citibank-Manila was frozen), among other factors, as reflected in its low Moody's rating of Ba-3, and Vanuatu is remote and largely unknown (except to Australian vacationers and readers of James Michener's novel *South Pacific*).

Another significant difference is the schedule chosen for initiating the individual elements of the package of offshore services to be offered. Labuan and Malta positioned offshore banking at the beginning of their implementing of offshore services, while Belize is timing it for a later implementation, qualified by the limited *de facto* offshore banking it has already permitted to develop.

Assessment of the Probable Future Development of the Three Eurobanking Centres

Macro-economic forecasting is obviously fraught with peril, leading to the observation that economists were invented to make weathermen look good. Nevertheless, the parameters affecting the future development of Labuan, Malta and Belize are sufficiently clear to venture some projections on the likely future of the three centres.

First, all three centres are pursuing well-thought-out strategies, but those of Labuan and Malta are significantly more sophisticated, reflecting the much stronger capabilities of their governments and central banks, and their larger, more advanced economies. Relatedly, Malaysia and Malta allocate significantly more resources than Belize to the global promotion of their offshore activities. Second, Labuan benefits from being in the high-growth South-East Asian region, while Belize too benefits from being in an economically strong area; conversely, Malta is handicapped to some degree by the slow growth of the Europe–North Africa region. Third, Labuan benefits from its relatively less competitive environment as compared with the older, highly sophisticated ones confronted by Malta and Belize. Finally, Labuan benefits greatly from the decision of the Malaysian government to further enhance Labuan by giving offshore banks there privileged access on a tax-exempt basis to the enormous and lucrative domestic Malaysian banking sector and economy.

Thus, while all three offshore centres are expected to continue to grow and prosper, the highest expectations are for Labuan, followed by Malta and Belize respectively. In particular, Kuala Lumpur's goal of turning the tiny island of Labuan into a 'mini-Singapore' is making good progress, but its very success may contain some seeds of its own undoing by encouraging other regional countries, such as Indonesia (which has 14,000 islands), to create new, competitive Eurobanking centres.

References and Further Reading

Bank for International Settlements, Monetary and Economic Department (1994) *International Banking and Financial Market Developments*, Basle: BIS.

Bank Negara Malaysia (1993) *Annual Report*.

Bank Negara Malaysia (1994) *Quarterly Bulletin*, vol. 9, no. 3.

Bank Negara Malaysia (1994) *Monthly Statistical Bulletin*, July.

Bank Negara Malaysia (1995) *Quarterly Bulletin*, vol. 10, no. 2.

Barrow, L.Y. and Williams, R.R.A. (1993) 'Belize.' In *Offshore Trust Yearbook*.

Belize Holding Inc. (1994) *Annual Report*.

Central Bank of Belize (1993) *Eleventh Annual Report and Accounts 1992*.

Central Bank of Belize, Research Department (1995) *Quarterly Review*, vol. 19, no. 1.

Central Bank of Malta (1993) *Twenty-Sixth Annual Report and Statement of Accounts*.

Central Bank of Malta (1994a) *Quarterly Review*, vol. 27, no. 1.

Central Bank of Malta (1994b) *Quarterly Review*, vol. 27, no. 2.

Central Bank of Malta (1995) *Quarterly Review*, vol. 28, no. 2.

Godrey, G.D. (1992) 'Belize – Open for Business.' (1992) *Offshore Investment: Journal for the Offshore Institute*.

International Monetary Fund (1994) *International Financial Statistics*.

Malaysia, Government of (1991) *Laws and Regulations in Respect of Labuan IOFC*.

Malaysia Industrial Development Authority (1993) *Malaysia: Your Profit Centre in Asia*, 1st edition.

Malaysia Ministry of Finance (1991) *Labuan: An International Offshore Financial Centre*. Kuala Lumpur: Malaysian Industrial Development Authority.

Malta Business Weekly (1995), various issues.

Malta Economic Update (1995), various issues.

Malta International Business Authority (1993) 'Malta.' In *Offshore Trust Yearbook*, pp. 74–79.

Miami Herald, International Edition (1994) 'Belize Fast Turning Into Offshore Financial Haven.' November: p. 1.

Sarver, E. (1990) *The Eurocurrency Market Handbook*, 2nd edition. New York: New York Institute of Finance (division of Simon & Schuster/Prentice-Hall).

Shearn Delamose & Co. (1993) 'Labuan.' In *Offshore Trust Yearbook*, pp. 216–221.

4

Banking, Finance and Offshore Banking Activities in the Island of Mauritius

Indur Ramphul

Mauritius: the Developing Financial Services Sector

It is widely acknowledged that among the major functions of the financial system the generation of an efficient allocation of resources, and ensuring the operation of a developed payments system to support economic growth in an environment of stable prices, are central. It is with these objectives in mind that during the 28 years of its existence, the Central Bank of Mauritius has played a decisive role in shaping the new strategic course of Mauritius's banking sector.

Mauritius is an island, 1,865 square kilometres in area, situated 2,400 kilometres off the eastern coast of Africa in the south-western Indian Ocean (see Figure 4.1). As recently as 25 years ago, Mauritius was largely a monocrop sugar economy. The banking infrastructure on the island hardly covered the urban areas, and banking transactions were limited to the raising of deposits for lending to the limited sectors of economic activity. The Central Bank played a major role in creating a conducive environment for banks to widen the range of their services as well as their branch network.

Today the island has an extensive banking coverage, with roughly one bank unit per 6,000 head of population. It possesses a fully fledged exchange economy, with banks providing a wide range of financial services which increasingly involve the use of information technology. There are 10 domestic banks, with a network of 150 branches operating in the island serving a population of just over 1 million inhabitants. The financial services system benefits from the support of 25 insurance companies and pension funds, a post office savings bank, and a National Mutual Fund as well as investment trusts. The banking industry of Mauritius has been extremely active in

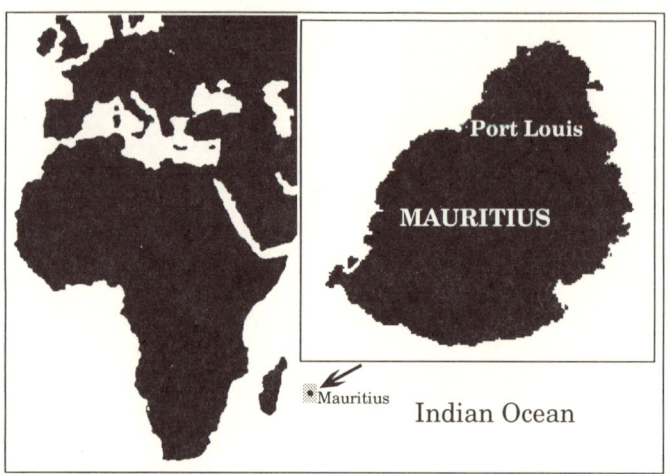

Figure 4.1   Location map of Mauritius

fostering the diversification of the economy into other sectors from its purely agricultural status in the 1960s. This diversification currently encompasses a textile manufacturing sector and a booming tourist industry. As a mobilizer of savings, the banking sector has intermediated effectively to support both the working capital and the investment needs of industry. This was necessary in the earlier years of economic development owing to the absence of an organized capital market in the economy.

Background: the Economic Development of Mauritius

In the context of world economic recession in the 1970s and the oil price shocks, from 1974 to 1975 Mauritius's balance of payments registered successive deficits which had to be tackled by the reinforcing of exchange control measures, the application of credit ceilings on banks and other quantitative controls such as the fixing of interest rates. The situation became so dramatic that at times the country had to manage with a few days' import equivalent of foreign exchange reserves. Commercial borrowing from international markets was soon reflected in high debt servicing requirements, particularly in an environment of rising levels of international interest rates. The lack of economic growth was also manifest in a rate of unemployment in the region of 20 per cent and rising inflation rates, as Mauritius's terms of trade deteriorated owing to a decline in the international price of sugar, the main export product of Mauritius at that time.

In an effort to redress this deteriorating situation, in 1979 Mauritius embarked on an economic stabilization and structural adjustment programme the effect of which was, *inter alia*, gradually to liberalize the banking and financial sectors.

Mauritius has always considered it essential to operate in close association with its markets, which are mainly located in Western Europe and North America. Thus, in 1974 there was a transition from the Commonwealth Sugar Agreement with the United Kingdom to the EU–ACP [African, Caribbean, Pacific] Lomé Convention. Under the convention, Mauritius strengthened its ties with the European Community by gaining duty-free access for its sugar exports. Similarly, notwithstanding the geographical remoteness of the island from the United States, Mauritius has regularly supplied a share of American sugar imports. In 1983, when the Preferential Trade Areas for Eastern and Southern Africa was formed, grouping a score of neighbouring African countries, Mauritius became a founder member of this upcoming economic bloc. It recently joined the Southern African Development Community (SADC), which groups a dozen Southern African states, including the Republic of South Africa. Discussions are continuing for the establishment of an Indian Ocean Rim Organization, of which Mauritius will be a member, along with other economies such as Australia, South Africa, Kenya and India. In this way, Mauritius has taken care not to stand isolated on the international economic scene. This policy approach has paid dividends.

As a consequence of Mauritius's remoteness from the main centres of economic activity, notably Europe, North America and the East, the island has fostered regular air and shipping links with the rest of the world, equipping itself at the same time with a very modern international communication infrastructure. All this underscores the necessary outward orientation of the economy. Had it not been for the wide-ranging access generated by facilities developed over the past two decades, Mauritius's export manufacturing activity based on imported raw materials would not have materialized. Nor would tourism have emerged as the third most important foreign exchange earner after manufacturing and sugar exports.

Implementation of the economic stabilization and adjustment programme proved extremely successful. After ten years of successive deficits, the balance of payments turned around as from 1984–85. Income tax revenue increased substantially, cutting down the fiscal deficit from its earlier level of 13 per cent of GDP to around 2 per cent currently. The country's debt service ratio has declined from around 20 per cent of GDP in the 1970s to less than 7 per cent, with all the obligations to the IMF and the commercial banks having been prepaid. Sound economic management has successively led to the improvement of Mauritius's international credit rating. The country's 40th rank (according to *Euromoney*, April 1992) among 162 countries is currently second only to Botswana as far as the African continent is concerned.

Fiscal and other incentives given to producers have also been enhanced, especially in favour of activities directed at the external sector. For example,

exporting companies were exempted from customs duty on import of equipment and raw materials, while several concessions in the form of low-cost
infrastructure, tax holidays and other facilities such as provision of cheap
working capital through the banking system were also made available. This
incentive regime was extended to the tourism sector, a natural area of
economic diversification for a richly endowed island environment. Gross
domestic product recorded high and sustained rates of real growth, rising to
12 per cent. The high rates of economic growth were spearheaded principally
by the performance of the export manufacturing sector, which undertook the
transformation of imported raw materials to service export markets located
in Europe and North America. By the mid-1980s the export manufacturing
sector overtook the sugar sector both as the main earner of foreign exchange
and in terms of the level of employment. The country's stock of foreign
exchange reserves began to increase, and since 1985 has averaged the
equivalent of 6 months' imports. Per capita income also increased markedly
during this period, and stands at present in the region of US$3,000. Full-
employment conditions have prevailed from 1987, leading to the importation
of labour services in the case of certain specific activities, notably textile
manufacturing and construction.

As part of the overall strategy to free Mauritius's financial sector from
counter-productive controls and restrictions affecting the efficient allocation
of financial resources, credit ceilings and limits on interest rates were gradually phased out. The process of liberalization of exchange control was
accelerated when in September 1993 Mauritius moved to IMF Article VIII
status. Exchange controls were completely abolished in June 1994. Currently
there are no restrictions on either current or capital account financial flows in
Mauritius.

With the abolition of all direct controls, there was a need for a shift in the
approach taken to the conduct of monetary policy, notably in the form of
indirect control. Specifically, the Central Bank proceeded with the auctioning
of Treasury Bills in 1991 in order to pave the way for the conduct of open-
market operations and the determination of interest rates on the basis of free
market forces. These initiatives have the principal objective of integrating the
functioning of the domestic financial market with the international markets,
while extending the scope for the free interplay of market forces.

The full-employment conditions prevailing since 1987 have also given rise
to a labour market reorientation. This has focused on the need for Mauritius
to move in the direction of higher value-added in production, as the continued shortage of cheap labour acted as a constraint on further economic
development. The focus of policy turned once again to internationally
oriented areas of activity.

At the same time, the development of the stock exchange of Mauritius,
which began in 1989, started easing to some extent the pressures exerted on
the banking system to provide funds for investment purposes. Some 50

companies, including a London-based company, are currently listed on the stock exchange. The number has gone on increasing alongside the 75 other companies whose shares are traded on the OTC market. Market capitalization has increased from $0.4 billion in 1992 to $1.7 billion in 1995. Five former public-sector companies are soon to be listed on the stock exchange. This is the result of an extensive privatization programme undertaken by the government with a view to widening the scope of the equity market. Moreover, foreign investors are now permitted to invest in the shares of companies listed on the stock exchange with the intention of promoting its internationalization and developing a new relationship with other regional equity markets.

Mauritius's comfortable foreign exchange reserves position and its relative economic prosperity thereby became the launching pads for the second phase of liberalization of the financial sector, which has responded positively to the policy initiatives underlying the liberalization of the economy. An active inter-bank money market is in operation, and financial institutions participate regularly in the weekly auctioning of bills. Liberalization policies have been matched by the consolidation of supervisory controls in the banking sector. Mauritian banks are required to observe the provisions of the Basle Minimum Capital Adequacy Standards. They are also required to adopt sound prudential policies in terms of concentration of lending risk and objective provisioning criteria for bad and doubtful debts. Their accountability and transparency to the public are accommodated by the requirement for annual publication of their accounts along the lines of International Accounting Standards 30 (IAS 30). Mauritian banks are set to face more competition in their foreign exchange dealings in view of licences to be issued to non-bank foreign exchange dealers operating in Mauritius. The aim is to heighten the skills and efficiency of the financial sector to enable it to continue competing on a level playing-field in the international arena, thereby enhancing the profile of Mauritius as a centre for international financial activity.

The Development of Offshore Banking Activities

Given Mauritius's strategic location in the Indian Ocean, its excellent communication links with the outside world, its outward-oriented policies and its long-standing background in international trade financing, the financial sector has been identified as the next area of economic diversification after industry and tourism. This strategy is appropriate in view of the resource constraints facing the economy.

Mauritian experience in manufacturing activity, based on the processing of imported raw materials mainly for export markets, has shown that a suitable policy package of incentives was a critical factor for access to international markets. Some of these considerations were judged equally relevant in the strategy to establish an offshore banking centre in Mauritius. Established financial centres, especially those with large and well-developed economic hinterlands, confer certain natural advantages for doing business as compared to emerging narrow-based economic locations. Taking the view that an aspiring financial centre has to attract business by somehow eroding such natural advantages of established centres, Mauritius has proceeded by putting in place a package of fiscal incentives for the development of its offshore banking sector. This process started in 1989 after relevant adjustments to the Banking Act.

Incentives provided by government to offshore entities with the specific aim of making Mauritius an attractive international financial centre in terms of cost competitiveness of operators included:

- an optional zero-rate tax on net profits arising from offshore business operations;
- free repatriation of profits;
- a concessionary personal income tax rate for expatriate staff at half the prevailing personal tax rate for residents (the maximum tax rate for residents is 30 per cent);
- complete exemption from taxes on imported office equipment;
- exemption from stamp duties on all documents relating to offshore business transactions;
- complete exemption from import duties on cars and household equipment for two expatriate staff per company;
- no withholding tax on interest payable on deposits raised from non-residents by offshore banks;
- no withholding tax on dividends and benefits payable by offshore entities;
- no estate duty or inheritance tax payable on the inheritance of shares in an offshore entity;
- no capital gains tax or other hidden tax.

In this context, along with the extensive domestic bank network, seven offshore banks have been licensed since 1989. This carries one step further Mauritius's interface with international markets, and gives a new direction for financial sector development. The offshore banks established in Mauritius's offshore financial centre are affiliates of international banks of high reputation, namely Barclays Bank plc, the Bank of Baroda, the Banque Nationale de Paris Intercontinentale, State Bank International (a joint venture of the State Bank of India and the State Bank of Mauritius Ltd), Banque Privée Edmond de Rothschild (Ocean Indien) Ltée, Banque Internationale des Mascareignes (a joint venture of Crédit Lyonnais of France, Banque de la Réunion and Mauritius Commercial Bank) and the Hongkong and Shanghai Banking Corporation.

Several factors favour investor access to offshore banking in Mauritius. A long history of banking and finance, the openness of the economy, respect for law and a ready supply of ancillary services, such as internationally trained lawyers, consultants and accountants, are among the important factors. There is also a long-standing history of political and social stability, with a formal guarantee against nationalization. Moreover, Mauritius has been involved in merchant trading since the colonial days, having played the role of a successful entrepôt for trade between the East and the West. Modern infrastructure and a positive business environment in the island are seen as the requisite tools for positioning the economy as a low-cost base for international operations.

This liberal economic policy outlook has led to substantial origination of investments for Mauritius's manufacturing sector from far-off geographic areas in the Far East, Europe and North America, thus establishing close economic ties with the developed markets. It has been said that the trade follows the flag. In the case of Mauritius, the presence of wide-ranging overseas investors, and the development of trading with their related markets, is viewed as a means for banks to intensify their financing activities by using the offshore banking sector as a convenient channel for routeing international financial business. Mauritius has positioned itself suitably in the evolving world economic order, in order to tap a share of the large financial flows which characterize activities in these developed markets. This approach explains the establishment of offshore banking in Mauritius and the success of this sector to date.

With a view to making Mauritius a regional financial centre, the legal framework governing offshore activities based in Mauritius was revamped in 1992, giving rise to the establishment of the Mauritius Offshore Business Activities Authority (MOBAA). MOBAA acts as a one-stop shop, acting as a co-ordinator between public-sector agencies and offshore units. This enabling legislative environment has been reinforced on several fronts, including the introduction of legislation relating to the establishment of international companies, modern offshore trusts and international funds. It is in this context of clearly projecting the role of Mauritius as an international offshore centre with an expanding business base that a freeport was also established in 1993 to facilitate transhipment and small-value-added activities for international trade directed at the regional economies.

The success of Mauritius's offshore activities is evident in the increasing numbers of non-banking offshore companies, of which there are over 5,000 currently. Moreover, the attractive regime governing freeport activities has successfully attracted to Mauritius some 400 operators in the freeport, some of which are affiliates of substantial international operators. The deepening of offshore banking activities based on this growing number of non-bank offshore companies located in the centre confirms, as it were, the completion of a first phase in the development of the international financial services sector of Mauritius in the seven-year span since offshore banking activity was introduced. It also marks the beginning of the concrete realization of the economic diversification plan underlying the initiation of the offshore financial sector.

Reasons for the Success of Offshore Activity in Mauritius

This success of offshore activity in Mauritius can be largely attributed to the conducive overall environment in Mauritius, and to the availability of a pool of qualified personnel servicing the offshore sector. The low costs of operation of offshore units represent another important environmental factor favouring this development. For example, offshore banks licensed in Mauritius are required to maintain capital amounting to approximately US$1.5 million, which need not be domiciled in Mauritius but may be freely invested in convertible currencies. Moreover, office space is not expensive: the cost of prime furnished office space in the city of Port Louis is in the range of US$14 to US$20 per square metre per month. Qualified clerks are available at about US$300 per month. The running costs of an offshore banking unit with five to seven staff members, including messaging of transactions, office rental payments, wages and similar recurrent bills, average some US$200,000 per annum. This suggests that the cost factor is seen as a competitive tool for generating profitable offshore business.

The government has adopted a proactive attitude at all times, taking care to initiate and support a suitable environment for the sustained development of offshore activities in the economy, and the generation of a necessary critical mass of activities for enhancing offshore sector development. It is not surprising, therefore, that Mauritian offshore banks have been able to tap business, including raising of non-bank deposits, from a wide spectrum of no fewer than 50 countries worldwide, including the well-developed financial centres. To extend the scope of offshore business, government is also actively engaged in the signing of a comprehensive network of double-taxation treaties. At present, Mauritius has signed such treaties with the United Kingdom, France, Germany, India, Italy, Zimbabwe, Swaziland, Malaysia, Sweden, the People's Republic of China, Madagascar, Pakistan, Namibia, Luxembourg and the Republic of South Africa. In particular, the double-taxation treaty with India has resulted in an appreciable volume of business involving non-resident Indians and other international investors undertaking fund management and investment business in India through the agency of entities operating within the Mauritian jurisdiction.

The following types of offshore activities can be conducted from Mauritius:

- offshore banking;
- trustee of offshore trusts;
- offshore insurance;
- offshore funds management;
- international financial services;
- operational headquarters;

- international assets management;
- shipping and ship management;
- international licensing and franchising;
- international data processing and other information technology services;
- offshore pension funds;
- international trading;
- international consultancy services;
- international employment services;
- aircraft financing and leasing.

The combined effects of this overall positive environment have proved beneficial in several respects. First, Mauritian offshore banks have become profitable within a short period after their establishment. Second, the range of offshore business undertaken from Mauritius has gone on expanding, thereby contributing to a significant enlargement of the base on which offshore business is conducted. These developments have confirmed that the policy initiative to diversify the economy on the basis of financial services directed at international markets has brought about the necessary comparative advantages to operators in Mauritius's offshore business centre. This process will strengthen as units established in Mauritius continue to interact with each other, thereby widening the business and skill base of the economy.

Mauritian offshore banks are licensed and supervised by the Bank of Mauritius. The licence also enables a bank to undertake trusteeship of offshore trusts, international fund management business and custodian services. Mauritius is a member of the Offshore Group of Banking Supervisors comprising 19 financial centres across the world, all of which are committed to effective banking supervision in accordance with international standards. The overall peaceful environment of the island, added to international recognition as a well-supervised centre, inspires the confidence of investors and bank customers. Another factor, namely banking secrecy, acting as a bedrock of confidentiality for customers of offshore banks, has also contributed to the success of the Mauritius offshore centre. A major objective of this overall strategy is to provide a sound offshore banking sector capable of attracting as customers of the offshore banks individuals of high net worth and corporates involved in international financial business.

In a sense, offshore activities may be seen as an avenue for expanding trade and other business network links with the rest of the world. Freeport activities are one example of such a model of development adopted by Mauritius, whereby the absorptive capacity of large economic entities in the surrounding region and worldwide is serviced from Mauritius with value-added gains. Through its membership of the 18-member, 225 million population economic bloc known as the Preferential Trade Area of Eastern and Southern Africa (PTA), now COMESA, Mauritius provides an important means for overcoming sovereign risk in trade-related payments. The COMESA Clearing House grouping all central banks of the COMESA region entails inter-central bank settlements for all trade transactions, a means for avoiding individual country risk associated with normal trade transactions, while also generating opportunities for expanding trade in the region. In this context, it is expected that the development of a thriving offshore banking sector in Mauritius should go hand in hand with the generation of favourable effects on the trade and economic growth of neighbouring countries.

The absence of withholding or other taxes on income of offshore banks or their customers acts as part of an attractive package to customers of offshore banks. It is also cost-effective for the offshore banks themselves, as no derived 'carrying costs' are imposed on their cash flows. Moreover, the seven offshore banks presently operating from Mauritius are affiliates of established international banks of high reputation. They consequently have the necessary skills to satisfy customer demands for sophisticated financial products, and also possess access to their parent company networks in diverse regions. When the regional economies, which are potentially rich, eventually develop to the level of thriving economies, the offshore banks already established in Mauritius stand advantageously poised to capture the resulting business. Over time, therefore, progress will be based on a wider area of international financial interaction, and the challenge for Mauritius's financial sector is to ensure that all genuine potential business remains within the competitive reach of Mauritius's financial agents.

The Way Ahead

Since 1989, when the first offshore bank was set up, Mauritius has registered considerable success in developing its offshore banking sector. The volume of business handled by offshore banks has gone on increasing from year to year, as has their range of activities and the geographical coverage of their business. The Mauritian authorities are keenly aware of the intensive competition entailed in operating the offshore banking sector and the corresponding need to sustain any competitive edge by pursuing the appropriate policies. Moreover, the authorities are anxious to preserve the sound reputation of Mauritius as a well-regulated financial centre, giving both the offshore bank operators and their customers the necessary confidence and financial advantage for transacting business in the local centre.

With the rapid development of international communications and the widespread use of information technology in the context of global finance, geographical remoteness or the smallness of a country is no longer a handicap. Although the notion of time zones is gradually being eliminated, with global 24-hour markets, Mauritius still holds a key intermediate position at GMT +4 between the Far East markets and the Western European markets. Mauritius is well aware that the extent to which it will be able to equalize competitive advantage – in terms of the environmental, financial and legal aspects of business – will determine how effective a player it will become as an offshore banking jurisdiction. What is more important, however, is both the ability to generate business in a cost-effective manner, thereby attracting the international investor, and the development of lasting customer relationships though the sustaining of its intermediating role by continuously raising new layers of expertise. Successful integration with related economies on a regional as well as global level must accompany this process.

Conclusion

A global financial environment without exchange controls implies that for any economy to interact successfully within the world financial marketplace, enhanced efficiency will remain the basic requirement. Moreover, financial institutions will be able to play their role effectively only if they are equipped with the latest technology and operate in an environment with the necessary financial experience and expertise.

Mauritius has forged ahead in the past years by removing all impediments to the effective development of its financial sector. The Central Bank has taken an active part in such development by encouraging financial institutions to employ state-of-the-art technology in a cost-effective manner. The latest development in this field is the introduction of an automated clearing and settlement system, to be adopted shortly by all financial institutions in Mauritius. This will enable faster communications with international financial markets. As in the case of the manufacturing sector, the financial sector will thrive provided the response time for concluding financial transactions in the future is shortened and comparative advantage in the markets is maintained by the operators. This could be one of the major conditions for islands and small states to join the mainstream of development in the international financial marketplace.

5

Mapping the Minefield: Theories of Island Offshore Finance Centres with Reference to Jersey

Mark P. Hampton

Since 1945, international trade and finance have become increasingly glob-alized, with the emergence of a truly international financial system. From the 1960s on, the financial system expanded massively, with the growth of an international network of electronically connected financial centres and the creation of entirely new financial instruments and markets such as the offshore Eurocurrency markets. These were initially based in London and saw phenomenal growth from around $500 million in 1959 to over $63.4 billion by 1969 (Owens, 1974; Mendelsohn, 1980). Associated with this new type of finance was the transformation of certain tax havens (often islands) into offshore finance centres (OFCs).

This chapter discusses why Jersey (see Figure 5.1) emerged as an OFC during the 1960s and whether it was purely the result of the 1960s environ-ment and the changes to international banking, especially in London, or whether there was more involved. First we examine the theoretical context and the main approaches towards offshore finance, discussing why Jersey emerged as an OFC. This is analysed over two phases: 1955–61 and 1962–71. We then evaluate the role of the different key factors in the emergence of the OFC and examine their significance in relation to each other and their changing hierarchy over time.

Theoretical Considerations

Before we discuss offshore finance further, it is helpful to start by defining what an OFC is:

> a centre that hosts financial activities that are separated from major regulating units (states) by geography and or by legislation. This may be a physical

separation, as in an island territory, or within a city such as London or the New York International Banking Facilities (IBFs). (Hampton, 1994, p. 237)

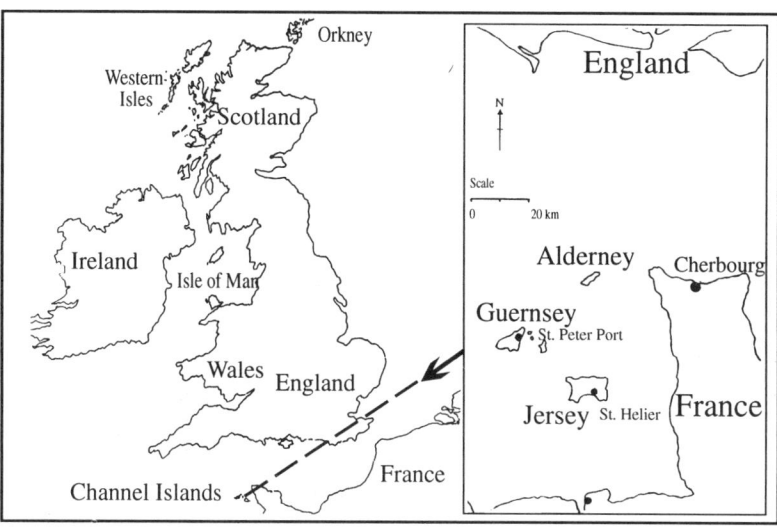

Figure 5.1 Location of the Channel Islands in the British Isles

Since we are concerned with Jersey we shall mainly be considering island OFCs, as these are the majority of OFC locations.[1] Table 5.1 illustrates one typology of island OFCs grouped by function. The table was constructed from the patchy comparative data available on island OFCs – the patchiness stemming from the basic secrecy of this industry. Nevertheless, it indicates that all the tier 1 functional OFCs are dependencies rather than independent nation-states. In tier 2, compound OFCs, the majority are independent nation-states. Of the total of 30 island OFCs, 13 are non-independent islands at present. Thus the majority of island OFCs are linked to the United Kingdom in some manner, either as dependencies, as colonies or as former territories. Two other island OFCs are linked to the Netherlands. Interestingly, none of the US, Danish or French territories are OFCs. A partial exception is Vanuatu, but as it was a joint British–French condominium until independence in 1980, there was at least some connection with the United Kingdom.

In comparison with other areas of economics, offshore finance is a somewhat problematic area for academic study. Essentially this is due to the nature of OFC activities, which are based upon secrecy and confidentiality. OFCs' customers demand secrecy from revenue authorities, governments and sometimes their own families. This criterion applies even for OFCs' legitimate business customers, the transnational corporations (TNCs) and

Table 5.1   Island OFCs grouped by function

Small island	Political status
Tier 1. Functional OFCs[a]	
Bermuda	UK colony
Isle of Man	UK dependency
Guernsey	UK dependency
Jersey	UK dependency
Tier 2. Compound OFCs[b]	
The Bahamas	Independent
Bahrain	Independent
Barbados	Independent
Cayman Islands	UK dependency
Cyprus	Independent
British Virgin Islands	UK dependency
Netherlands Antilles	Netherlands dependency
Vanuatu	Independent
Tier 3. Notional OFCs[c]	
Anguilla	UK dependency
Antigua and Barbuda	Independent
Aruba	Netherlands dependency
Cape Verde	Independent
Cook Islands	Independent
Grenada	Independent
Labuan	Malaysian federal state
Malta	Independent
Marshall Islands	US associated state
Montserrat	UK dependency
Nauru	Independent
St Kitts and Nevis	Independent
St Lucia	Independent
St Vincent	Independent
Tonga	Independent
Turks and Caicos	UK dependency
Western Samoa	Independent

[a] A functional OFC has actual financial activity taking place, the location of full branches of banks, plus other services (fund managers, trust companies, etc.). The OFC employs over 12 per cent of the labour force and contributes over 25 per cent of GDP.
[b] A compound OFC has a mixture of functional and notional activities, perhaps being a centre with an increasing number of shell offices becoming fully operational branches. The OFC employs 3–11 per cent of the labour force and contributes 10–24 per cent of GDP.
[c] A notional OFC has brass-plate or 'cubicle' offices of banks making book entries of transactions. The OFC employs under 3 per cent of the labour force and contributes under 10 per cent of GDP.

wealthy individuals who seek confidentiality. Additionally, two further aspects can hinder data collection: first, the various OFC host governments' attempts continually to portray themselves in the best possible light, and second, the fact that financial services are fundamentally based on the fragile notion of 'confidence'. This results in all players wanting to promote the respectability of their OFC. In combination with the host government's stance, this can lead to reluctance to reveal information about offshore activities. Therefore, this results in a lack of international comparative data on different OFCs.

Recent research includes the work of the Economist Intelligence Unit (Doggart, 1993) and the writings of tax lawyers and other practitioners (Langer, 1988; Spitz, 1990; Diamond and Diamond, 1990). However, most are technical works, written for international tax specialists and generally lacking conceptual depth. Overall, offshore finance is an under-researched area, possibly because of the lack of data and the problems noted above. The few academic studies on offshore finance are R.A. Johns's monographs (Johns, 1983; Johns and Le Marchant, 1993) and the works of Walter (1985), Naylor (1987) and Ehrenfeld (1992). Johns approaches offshore finance from a broadly neoclassical viewpoint, and therefore provides a useful challenge to alternative theories of offshore finance. His helpful contributions will be further discussed later.

Walter, Naylor and Ehrenfeld all focus on the criminal abuses of offshore finance. This group could be dubbed the 'secret–hot–evil money' approach, perhaps a reflection of the marketability of popular works on the sleazier aspect of international financial flows. Despite some useful comments on the development of particular OFCs, all three authors lack any conceptual framework to explain why offshore finance emerged.

Next we need to consider offshore finance in relation to wider theories of international banking, the behaviour of the TNC and international finance centres. However, there is one caveat: OFCs are not just sites of international banking, but may host a variety of financial services such as offshore trusts, private companies, offshore funds and captive insurance so that we may not be able to graft OFCs on to the main body of international banking theory. Similarly, this applies to most of the theoretical work on international finance centres (Reed, 1981; Park and Zwick, 1985; Choi *et al.*, 1986; Sassen, 1991; Coakley, 1992; Thrift, 1994), because OFCs are often located in microstates rather than in global cities. For instance, most OFCs are not the location of the headquarters of international banks (a common criterion used to rank financial centres). OFC banks are often branches or subsidiaries of onshore banks, so that at present very few of the top 200 banks as measured by assets are headquartered offshore.[2]

Having noted this, we can broadly categorize two major approaches or world-views of offshore finance: that of orthodox economics and that of alternative, generally Marxist-based, approaches. Within the orthodox economic approaches, the central discussion pivots around the concept of the free market and whether it is perfect or imperfect in its operations. If markets are perfect (the neoclassical view), then there is less need for government intervention, resulting in an emphasis on deregulation. The 1979 suspension of UK exchange controls by the incoming Thatcher administration is a good example of this tendency, and was further developed into wider deregulation of the City of London in the 1980s. In this approach, the emergence of offshore finance is seen as a rational response to the constraints of the nation-state, exemplified by onerous onshore regulations.

If, on the other hand, markets are imperfect in their operations (the Keynesian view), then government intervention is necessary to regulate the markets efficiently. Thus, in this view, offshore finance has emerged because of a lack of effective international regulation, and so the policy prescription is for more effective onshore regulation.

The first orthodox view, that of the perfect market being distorted by government action, is a common theme in much of the literature on offshore finance: 'Banks did not invent the Euromarket. Governments created it by seeking to control the natural flow of money' (Theobald, 1981, p. 19).

Johns describes the creation of an onshore 'international friction matrix' (1983, p. 8) of bank regulation and taxes that distort the operations of free markets so that funds flow to 'zero friction' (1983, p. 18) offshore centres. This view is also shared by Park (1989), Doggart (1993), Walter (1985) and Naylor (1987). The last two authors focus more on the abuse of OFCs and illegal activities, although they concur that onshore regulation and taxation lead to increasing 'underground' economic activities, including offshore transactions. Lessard and Williamson (1987) comment on the proactive role of offshore banks in deposit-seeking from onshore sources.

If we consider the imperfect-market orthodoxy, we can track the rise of this approach since the 1960s in the writings on the TNC. The early work of Hymer (1960) on industrial organization and firms' behaviour stimulated many later theories including the 'internalization' school exemplified by Buckley and Casson (1976). Aliber (1976) argued that the international expansion of banking was linked to domestic bank concentration where high profits would lead to overseas expansion. However, this was challenged by Cho (1985), who contended that a low concentration of banking drove banks overseas to avoid competition.

Grubel's (1977) 'three-stream theory' identified three types of multinational banking. Of these, two are of interest here: multinational service banking and multinational wholesale banking (international lending). Although both of these may take place in OFCs, Grubel argues that the two are driven by different motives. The former happens where banks follow TNC clients, whereas the latter arises from government-induced 'imperfections' in international capital markets. This idea was further developed by Giddy (1983), who argued that the form and location of international banking were due to market imperfections and that regulation led to monopolistic profits. Thus if there were not differential regulations or other market imperfections, all banking would be at arm's length.

However, we can criticize all these orthodox approaches as being somewhat superficial. Essentially, their main argument is that offshore finance can be ascribed to onshore regulation and taxation that push business offshore. Although the two main wings of orthodox economics argue vociferously with each other about whether the market is perfect or imperfect, our understanding of offshore finance has not really been advanced. In short, the explanation offered is only one small step further than the 'key-factors' approach set out in many OFC banks' promotional literature.

Therefore, to develop the analysis of OFCs we need to go beyond this. We could attempt to group key factors, or examine whether there is a hierarchy among them and whether that changes over time. However, all the key factors of taxation, regulation and political stability operate within a historically specific context of certain political relationships. This can be illustrated if we compare two British islands, Jersey and the Isle of Wight. The fundamental reason why the former is an OFC and the latter is not is clearly a function of the nature of its political relationship to the United Kingdom. The Isle of Wight is a small British county, while Jersey is a Crown dependency and enjoys a large measure of political autonomy from the United Kingdom. This results in freedom for the Jersey government to set its own tax rates, banking and other commercial regulation to attract both funds and OFC firms. In contrast, the Isle of Wight is not in a position to do this as it falls under the full jurisdiction of UK law.[3]

Most present OFCs are either dependencies, ex-dependencies or colonies of a major power and have had some relationship with a mainland country. Examples include the Channel Islands, the Isle of Man, the Netherlands Antilles and the Cayman Islands. As such, the peculiar relationship of being 'within and yet without' the onshore country sets the context for offshore development. Therefore, an explicit focus on the political economy of the onshore country will advance our discussion of offshore finance and help explain why certain policies have been followed towards international banking. This approach helps comprehension of why the UK government encouraged London's development as an international finance centre in the 1960s. In other words, we need to progress to a different level of analysis from the work discussed above. This leads us to consider an alternative approach that incorporates the elements of political economy.

An alternative to orthodox economics is to use a materialist-based approach, built initially upon the work of Marx and then further developed by later writers. This broad Marxist approach argues that offshore finance can be seen as the outcome of the conflict between state and capital. The conceptual tool of the fractions of capital is used to explain the differing interests of industry and finance in the onshore economy: 'In the twentieth century, some aspects of politics can be understood as a reflection of the struggle between finance and industrial capital' (Cole *et al.*, 1983, p. 252). Thus the relative positions of industrial and financial capital in the onshore economy may be of great significance (Coakley and Harris, 1983). At this highly abstract level, these two fractions of capital compete with each other for a share in the surplus value created, whether as profits or as interest. Coakley and Harris argue that the historical dominance of financial capital over industrial capital in the United Kingdom has long been associated with the City of London as a major finance centre.

What is meant by the term 'financial capital' as distinct from 'finance capital'? Hilferding (1981) developed the concept of 'finance capital' to analyse capitalism in the early twentieth century, particularly the process of concentration of banking and the integration of banks and industrial companies. Lenin described this as a 'personal union' (1965, p. 45) between the banks and the largest industrial concerns. But, as Coakley and Harris (1983) point out, Hilferding's use of the term may not be entirely apposite for capitalism in the late twentieth century as it refers to the integration of financial and productive capital. While this was the case in the United States (in the first decade of the twentieth century) and in Germany, the growth of TNCs with the internationalization of both productive and financial capital makes the classical definition somewhat problematic. Here 'financial capital' refers to a fraction of capital rather than Hilferding's definition.

Financial capital has certain characteristics, particularly that it is highly mobile and liquid, unlike industrial capital. This means that financial capital requires a certain amount of freedom to operate without the constraint of onerous government regulation. Ironically, as elsewhere under capitalism, a complete lack of any regulation would also be fatal to financial capital as chaos would ensue from the unleashed competition between capitals.

Thus, we can posit that the relative freedom of financial capital in the United Kingdom resulted in the development of offshore finance in the 1960s as manifested in the unregulated offshore markets (Eurocurrencies) and the associated creation of OFCs. Both innovations were highly profitable for financial capital. Eurocurrencies were highly profitable because the lack of reserve requirements for banks created generous margins (Grubel, 1982). For similar reasons, operations in OFCs were also lucrative, as OFC offices are often highly profitable (States of Jersey, 1993, pp. 49–50).

Thus it is reasonable to suggest that the relative freedom of financial capital in relation to industrial capital is significant not just for mainland economies, but also for OFCs. If we consider an onshore economy with relatively weak financial capital (subordinate to industrial capital), it is reasonable to assume that the overall effect will be that the country is less likely to tolerate or encourage nearby OFCs. It is not in the interest of industrial capital to allow such freedom for financial capital. In the struggle between these fractions of capital, industrial capital, if dominant, will act against financial capital to constrain it, as financial capital, unhindered, has a propensity to expand and increase its share of surplus value.

This is *not* to suggest a simplistic causation that strong, or even hegemonic, financial capital equals a government that allows offshore finance, and weak financial capital the converse. However, it seems reasonable to suggest that the existence of a powerful and relatively free fraction of capital that is close to central government would be indicative of conditions that are more favourable for the growth of offshore finance. The opposite case, of relatively weak and constrained financial capital and relatively powerful industrial capital, would mean that the conditions for offshore finance would be less favourable. Additionally, in this second scenario, financial capital is more distant from central government and therefore less able to influence decisions that might allow it the fiscal and regulatory 'space' to develop offshore finance.

Although there might exist parts of the state that would want to control any OFC – such as the judiciary or administration – there might also be a general lack of interest from other parts of the state. The relative strength or weakness of financial capital could thus indirectly affect the government and administration's attitude, resulting in at least acquiescence to, if not tolerance of, nearby OFCs. Thus it is a more subtle hypothesis than a simplistic line of causation, but it seems to be a more accurate model to describe the British government's attitude towards offshore finance than the two orthodoxies detailed earlier.

The British government, indirectly via the Bank of England, permitted the establishment of the London Euromarkets in the 1960s, and the concomitant development of offshore finance in Jersey and elsewhere. This raises the question of why this was permitted, necessitating an exploration of the relationship between UK financial capital and the state. Several authors (Coakley and Harris, 1983; Ingham, 1984; Hilton, 1987; Cain and Hopkins, 1987) have stressed the power of UK financial capital, noting its proximity to, and influence over, central government. To examine this more rigorously, it is useful to separate this analysis into two parts: first, the issue of the relationship with the state, and second, the issue of the position of financial capital relative to other fractions of capital.

UK financial capital and the state

In the United Kingdom, finance and banking are often considered to be synonymous with the City of London, and in that sense the City can be used as shorthand for financial capital (Harris, 1985, p. 17). As in any economy, however, the interests of financial capital may coincide with some government policies, or they may conflict with those policies. In addition to actual policy, the impact of financial capital upon a given society or state is more than the sum of economic activity in that sector. As Hilton comments, 'The City establishment is not just a question of blood, it is a state of mind, a whole way of doing business, a complete approach to the world' (1987, p. 15). Nevertheless, the ideological impact of the banking world-view is a reflection of the underlying economic structure. In more concrete terms, the political power of UK financial capital is associated with the peculiar role of the Bank of England, which, although part of the UK state, also represents the interests of financial capital. As an organ of the state, the Bank can be directed by the Treasury to follow government policy to lead the financial sector. On the other hand, the Bank also acts in the interest of financial capital to influence government policy. Examples of the latter include the return to gold in 1925, government spending cuts in 1931, and the 1967 devaluation of the pound (Pollard, 1969; Coakley and Harris, 1983; Williamson, 1984).

In this complex triangular relationship between the City, the Bank of England and the Treasury, much depends on the relative strength of each party. However, although the exact connection between financial capital and central government is hard to quantify, the UK relationship appears closer than that of US financial capital to central government (Hampton, 1996). In addition, the relationship of UK financial capital via the Bank of England with central government was also reflected in the official encouragement of London's role as an international finance centre in the early 1960s. Arguably, the Bank of England was successful 'in persuading the government to leave international banking in London less regulated than in many other financial centres and thereby encourage its growth as the world centre for Eurodollar banking' (Coakley and Harris, 1983, p. 27). Thus, this influential relationship gave UK financial capital the space to innovate and create the offshore currency markets and, later, OFCs.

The power of financial capital

If we consider the relative position of UK financial capital, it seems to have become the dominant fraction in the United Kingdom. In the struggle between the fractions of capital, UK financial capital was not subordinated to industrial capital and it avoided the fragmentation that its US counterpart suffered in the 1930s with the draconian Glass–Steagall Act and in the 1960s with various federal regulations. There are three reasons for this.

First, UK financial capital had become increasingly concentrated even by the first decade of the twentieth century. In 1914 there were around 20 joint-stock banks with more than 100 branches each in England and Wales, and in Scotland a mere eight banks with around 1,200 branches between them (Pollard, 1969, p. 14). In the United States, the comparable figure was in the thousands (Myers, 1970). Second, in the 1959 Radcliffe Report on UK banking the focus was upon interest rate issues rather than an examination of the clearing banks' oligopoly. Thus UK financial capital survived this major investigation without any serious curtailing of its activities by the state. Third, in the struggle between financial and industrial capital, UK financial capital had successfully expanded overseas in addition to its domestic operations. In terms of the struggle between capitals, industrial capital was unable to contain it, and in that sense was the loser in the competition for the share of surplus value.

London had a history of being an entrepôt with banking closely linked to foreign trade. Trade-dependent merchant capital was in a sense the fore-runner of the expansion of UK financial capital in the nineteenth century. The expansion of the British Empire and the opening up of new markets created a great demand for capital, so that London became the centre for the export of British capital both as portfolio investment (in stocks and shares) and as direct foreign investment.

Pollard (1969) argues that there was increasing separation of financial from industrial capital, with banks increasingly involved in speculation on the expanding stock-market or in overseas investment while not providing credit for UK industry. However, this view is open to debate. Coakley and Harris argue that rather than the simplistic model of the two fractions of UK capital (industrial and financial) being in straight competition, the reality was rather of an 'overseas alliance' between the two capitals. That is, financial capital became the overseas operation of UK industrial capital.

Given the changing positions in the continuing struggle between the two fractions, it is plausible that the alliance phase was but one part of the historical process that would eventually see financial capital as the dominant UK fraction. Coakley and Harris (1983), acknowledging this shifting dynamic, have suggested that the operations of financial capital since the 1960s represent an 'enclave' phase, one of internationally oriented activities which forms part of a global network of financial capital. Alternatively, this phase could be called 'offshore' operations. Thus, UK financial capital has become the dominant fraction of capital through first forging an overseas alliance with UK industrial capital at the end of the nineteenth century and then subsequently breaking free at the start of a truly international phase beginning in the 1960s. This resulted in the City (financial capital) becoming virtually an enclave within the UK economy; that is, effectively 'offshore'.

The existence of this enclave with its minimal regulation allowed increasing secondary bank activity and the development of the Eurocurrency markets. In turn, this eventually resulted in the expansion of UK financial capital to the new offshore (literally) environment of various low-tax islands which became OFCs. Thus, the British island OFCs such as Jersey are arguably the eventual, concrete results of the relative power and position of UK financial capital.

Jersey Case Study, 1955–71

In 1969 Jersey had an estimated 45 per cent of its GDP dependent upon tourist expenditure, 15 per cent from the finance sector and 25 per cent from the expenditure of wealthy residents. However, by the mid-1970s the OFC was playing an increasingly important role in the economy. Jersey was no longer a mere tax haven but had become a global-level OFC, hosting a variety of financial services including private banking, offshore funds, trusts and offshore companies. This development will be examined using the fractions of capital and the political economy approach.

The central theme in this case study is how the specific political relationship with the United Kingdom (and the relative power of financial capital within the UK economy) underpins the more obvious 'key factors' found in much of the offshore finance literature: regulation, taxation, secrecy, customer preference, political stability, location, technology, tourism and the labour force. This will shed light upon whether all these factors operate in the Jersey case, whether certain factors were dominant at different times and whether or not there was a rigid hierarchy of factors in operation.

Clearly, some factors such as regulation or taxation can be manipulated by governments, such as the 1981 creation of International Banking Facilities in New York. Other factors, such as location, are fixed. In the list of factors, regulation and taxation are arguably dominant, while secrecy, political stability and client preference are somewhat lower in the hierarchy and may be called supplementary factors. The remainder (location, technology, tourism and the labour force) may be seen more as prerequisites. This can be illustrated if we consider a counter-factual case of non-development of an OFC such as the earlier example of the Isle of Wight. Here, the island had all the prerequisites and political stability. However – and this is the crux of the argument – the twin dominant factors of taxation and bank regulation were not in operation as the island falls under full UK jurisdiction as an integral part of the UK. This brings us back to the underlying political relationship of island and mainland, and the political economy approach. To examine this further we can divide the Jersey case study into two phases:

- phase 1: notional tax haven (1955–61);
- phase 2: functional tax haven and sterling OFC (1962–71).

Phase 1: notional tax haven 1955–61

Taxation and wealthy immigrants

In this first phase there were two closely linked main elements: the arrival of wealthy immigrants in the island, and the concomitant expansion of financial services. Jersey started to attract numbers of wealthy immigrants: returning UK expatriates and UK high net wealth individuals (HNWIs). Post-war decolonialization had resulted in political instability in many countries as the British Empire collapsed. Therefore, many wealthy UK expatriates in areas such as east Africa were forced to reconsider their position, as returning to the United Kingdom would expose their funds to its high direct tax rates. Moving to Jersey became very attractive because of its low direct tax rate of 20 per cent. Jersey was also alluring to a second group of HNWIs, the UK *nouveaux riches* who were making large profits in the post-war boom, and wanted a low-tax haven for their funds and a warm, sunny place to live. For both these types of wealthy immigrants Jersey seemed to be ideal, with tax as the central consideration.

In 1958 local company law was modified so that Jersey companies could be used for explicit tax avoidance by UK residents and, increasingly, by the new HNWI arrivals in the island.[4] The simple mechanism for the latter group was to form a Jersey company to reinvest in the UK. The Jersey company was non-resident for UK tax purposes and thus the HNWI as the beneficiary owner avoided UK estate duties and, later, capital gains tax. This illustrates the importance of tax as the dominant factor during this phase.

Bank regulation

In 1955 the Jersey banking sector consisted of six banks – four UK clearing banks, the Jersey Savings Bank and one small local bank – which were subject to UK bank regulation, especially concerning capital flows and exchange controls. By the late 1950s and early 1960s there was a large flow of capital into the island. Bank deposits in Jersey increased from £32 million in 1958 to £44 million by 1961 and fed into the amount of local advances (loans and overdrafts), which rose by a massive 85 per cent between 1958 and 1960 (Powell, 1971, p. 151). However, high interest rates do not explain the rise. An ancient anti-usury law, the Code of 1771, which operated until 1962, imposed a local interest rate ceiling of 5 per cent p.a. on all banks including the Jersey branches of the UK clearing banks. With the UK Bank Rate rising above 5 per cent p.a. during the later 1950s and early 1960s, all else being equal, an outflow of capital from Jersey into the United Kingdom would have been expected. However, this did not appear to happen, there being no noticeable outflow of funds from the island in September 1957, for instance, when the UK bank rate rose above 7 per cent p.a.

Given that on interest rate differentials alone, it would have been irrational for investors to deposit funds in the island, other factors, including taxation, must have influenced the choice of Jersey as a location for these deposits. During the late 1950s and early 1960s, 70 per cent of deposits in Jersey originated from non-residents (Powell, 1971, p. 59).[5] Although some of these were simply relocating funds to Jersey prior to their moving house to the island, others were taking advantage of Jersey deposits as a UK tax loophole. In the late 1950s lawyers discovered a loophole in UK estate duty, where property held abroad by UK residents was not liable. Jersey, for the purposes of the law, was considered 'abroad' and so an estimated £12 million flowed into Jersey[6] 'as funds queued to get into the island' (Rutherford, 1976, p. 87). The successful manipulation of this UK tax loophole demonstrates that tax avoidance was more significant for the emerging OFC than the regulatory constraint of the interest rate ceiling.

The central consideration of the UK HNWIs, whether by transferring their funds to Jersey or as part of the process of physically moving to the island, was tax avoidance. Jersey's tax position, however, pivoted around the political relationship of island dependency with the mainland United Kingdom. The activities of UK HNWIs therefore illustrates clever exploitation of the peculiarities of this relationship, including the UK authorities' definition of the Channel Islands as 'abroad' for estate duty purposes. Thus for this first phase, tax was the dominant factor, although it can be argued that Jersey's low-tax status was inherently due to its unusual political relationship with the UK and the resultant local autonomy.

Phase 2: Functional Tax Haven and Sterling OFC, 1962–71

In Phase 2, the island saw a rapid expansion of the OFC as measured both by bank deposits and by numbers of banks in operation (Table 5.2). Between 1962 and 1967 several UK secondary banks[7] started to open Jersey offices. Hill Samuel arrived in 1961, Kleinwort Benson in 1962, and significantly, a US bank, First National City Bank, in 1968. Other banks followed, so that by the end of 1971 there were 30 banks in the island.

Table 5.2   Bank deposits in Jersey 1962–71

Year	Total deposits (£ millions)	Number of banks in operation
1960	39	7
1961	44	8
1962	50	12
1963	55	13
1964	61	15
1965	69	16
1966	113	21
1967	159	24
1968	208	27
1969	296	28
1970	436	30
1971	470	30

Source: Hampton (1996)

UK bank regulation

The penetration of UK financial capital offshore into Jersey took place at the same time as the rapid growth of the London offshore (Eurocurrency) markets, arguably a function of the uneven Bank of England regulation of UK banks. This unevenness allowed the unchecked expansion of a less regulated, secondary banking sector, reinforcing the abstract argument that UK financial capital was free to operate in a broadly beneficial environment.

Jersey banks came under the regulatory responsibility of both the UK and the Jersey authorities. Under the 1947 Exchange Control Act Jersey banks were treated as UK banks and thus came under Bank of England monitoring.

Banks wishing to deal in foreign currency needed Bank of England Authorized Depository status. In addition, Bank of England supervision staff paid monthly visits to the island.[8]

In Jersey, as in the United Kingdom, the differing weight of regulation between clearers and secondary banks affected bank activities, particularly lending. Bank of England credit restrictions applied to the branches of the clearing banks and to the subsidiaries of the largest merchant banks but not to the other secondary banks. Thus the differential regulation across the sector permitted Jersey subsidiaries of secondary banks, particularly UK finance houses and small merchant banks, to enjoy freedom from lending restrictions. The deposit figures for secondary banks in Jersey (unavailable before 1966) show a massive increase in comparison with the clearing banks. In 1966 there were deposits in the 'other deposit-taking institutions' of £36 million, roughly 75 per cent of the deposits held in clearing banks (£48 million). In 1968 deposits were broadly twice those of the clearers (£118 million as against £55 million) and by 1970 had increased to over four times as much (£325 million compared with £70 million) (Powell, 1971).

On the other side of the balance sheet, between 1965 and 1969 Jersey clearing bank advances increased by only 1 per cent, reflecting the Bank of England's credit restrictions. Again, the secondary banks' Jersey subsidiaries were able to lend without restriction both locally, and increasingly to the banks' UK parents. This flow of funds can be modelled as follows.

UK credit restrictions between 1965 and 1969 squeezed UK borrowers, who were then forced to look elsewhere for funds. This coincided with Jersey's net surplus of funds, where the supply of money in the island's banks exceeded local demand for loans even at the very low interest rates prevailing. As we have already seen, this large pool of funds was due mainly to Jersey's low-tax position.

The island had attracted both HNWI immigrants and also large deposits from UK resident HNWIs wanting to avoid estate duty. For the former there was a large tax advantage in selling their UK equities (thus avoiding the 41.25 per cent UK withholding tax) and placing the proceeds in a Jersey bank account.

Between 1964 and 1966, three UK secondary bank groups formed Jersey subsidiaries to collect these deposits; they were the Julian S. Hodge group, United Dominions and Whyte, Gasc & Co.[9] They could then lend their surplus funds back to their UK parents as short-term loans. By 1969, of the £190 million in assets in Jersey banks, £117 million was on deposit with parent banks in the UK, either on short-term loan or 'at call'. Another £43 million was placed at call with other sources, most likely other banks in the group in the UK or possibly Guernsey (Powell, 1971, p. 166). This offshore mechanism to bypass UK credit restrictions can be seen as a British microcosm of how US banks circumvented US domestic credit restrictions by going offshore to the London Euromarkets from 1966 onwards (Bell, 1973).

This illustrates that initially regulation was the dominant factor for the secondary banks' move to Jersey. However, the interaction with taxation, which had drawn the wealthy residents and their funds, is also clear. The secondary banks were able to exploit this surplus money in Jersey and lend to UK borrowers. For other banks, such as the UK clearers, regulation was less significant than tax. The clearers formed Jersey subsidiaries alongside their existing branches in the late 1960s (e.g. Williams & Glyns in 1966, Midland Bank Trust Corporation (Jersey) Ltd and National Westminster Finance (CI) Ltd, both in 1967). These were started to expand the private banking and trust operations, which had increased significantly since the 1950s, and to collect deposits from non-residents.

Jersey bank regulation

Four main areas of Jersey bank regulation influenced developments between 1962 and 1971: the removal of the interest rate ceiling; a lack of local reserve requirements; ease of entry for new banks; and, until the 1967 Depositors and Investors Law, generally minimal banking legislation. In 1962 the Code of 1771 was repealed, reportedly in part because of the encouragement of Lord Bearsted of Hill Samuel, who wanted to set up in the island (*Offshore Finance Magazine*, 1991). This removed the interest rate ceiling which then allowed profitable margins on bank lending. Banks operating in Jersey did not have reserve requirements imposed by the States of Jersey's Finance and Economics Committee (FEC), which supervised the OFC, although they needed an informal 'letter of comfort' from their parent bank. The letter was often brief with no mention of capital adequacy, and was not a legally binding guarantee, rather a 'moral obligation'.[10] A lack of reserve requirements can result in significant profit margins for banks (Grubel, 1982)[11] so that banks in Jersey could match more precisely the maturities of liabilities and assets and therefore hold minimal reserves. Thus they could both operate on keen margins and make high profits on their Jersey activities.

The Bank of England played an indirect role, as financial capital started to penetrate the island. When banks intended to start Jersey subsidiaries they kept the Bank informed of their intentions, and there was ease of entry for certain banks if they were subsidiaries of large UK parents or Bank of England 'authorized' banks. Officially, the States of Jersey, through the FEC, were responsible for banking in the island, yet the Bank of England still kept a close eye on events:

> 'the Jersey authorities have always regarded themselves as quite independent. But behind the scenes I think there is always close understanding and a certain amount of liaison . . . in those early days, and here I refer to the 1950s and 1960s, we would have been in constant touch with the Bank of England on an informal basis.' (Jersey interviews)

This echoes the strange relationship of 'within and yet without', of being under the UK sovereign umbrella and yet with the space to have a surprising amount of freedom. Also it perhaps indicates the close relationship of UK financial capital via the Bank of England to central government. This argument of benevolent UK state interest, or at least acquiescence towards OFCs (Hampton, 1996), is, unsurprisingly, strongly denied by Powell, Jersey's Chief Adviser:

> I disagree entirely with the view expressed that in some way the offshore finance centre in Jersey grew up under the protective wing of a benevolent Bank of England ... [and] am not aware of any noticeable influence on the island's finance centre development ... due to the actions of the Bank. (Letter to the author, 2 April 1993)

By 1967 the number of banks in Jersey had increased to 21 and the States of Jersey enacted the Depositors and Investors (Prevention of Fraud) (Jersey) Law. This required banks to submit annual returns to the States and generally there was a tightening of restrictions on banking activities. However, despite this basic regulation in 1970 there were two failures of locally owned banks, the Guarantee Trust of Jersey and the Walford Banking Company. Arguably, the local effect of these failures was to reinforce the FEC view that only banks of reputable parentage should be admitted to the OFC (*Offshore Finance Magazine*, 1991).

Taxation

Taxation affected different sectors of the emerging OFC in various ways during this period. We have already noted the effects of tax in combination with regulation for some secondary banks. For other banks, such as the merchants and trust departments of the clearers, the main activity was private banking for HNWIs. This mainly involved forming trusts and offshore companies to avoid taxes, but also utilized Jersey bank secrecy. In 1961 the UK government closed the estate duty loophole to stop the '*embarras de richesse*' (Senator Cyril Le Marquand, quoted in Rutherford, 1976).[12]

In Phase One we saw how the new, wealthy residents in Jersey could avoid UK taxes such as the newly introduced capital gains tax (CGT) brought in by the Labour government in 1965. They could realize UK assets as bank deposits in Jersey, which were then only liable for the local 20 per cent tax rate when counted as income. In addition, offshore companies were increasingly used to avoid UK taxes. UK resident HNWIs could form a Jersey offshore company to invest in the United Kingdom. The offshore company was only liable for a small fixed fee of £100 p.a. rather than any income tax. The number of offshore companies formed in Jersey rose swiftly from 17 in 1962 to over 470 by 1970 (Table 5.3). Some of these companies were offshore funds, which were a significant 1960s innovation.

Jersey company law did not allow the capital base of a company to be changed, but in 1965 two Jersey accountants created the concept of the

Table 5.3   Corporation tax companies and offshore funds in Jersey 1962–71 (totals)

Year	Corporation tax companies	Offshore funds (all types)
1962	17	3[a]
1963	25	3
1964	66	3
1965	110	5
1966	172	5
1967	210	6
1968	282	7
1969	400	11
1970	472	14
1971	571	15

Source: Johns (1983, p. 113); and data from Commercial Relations Office
[a]These three were closed-ended offshore funds – that is, traditional investment trusts set up before 1965 – as opposed to the later innovation of open-ended offshore funds

redeemable preference 'penny shares' to get around this restriction. Ebor (later Save and Prosper) launched the offshore fund, an open-ended investment company, in which the management company could redeem shares if more shareholders sold shares than bought them. Thus the fund's capital could be increased or decreased according to demand. The new offshore funds were used by Jersey HNWIs to invest in the United Kingdom to avoid estate duty and CGT. Secrecy was another major advantage, as investors could remain anonymous. Four funds were launched for investment in UK equities between 1965 and 1968, and another four between 1970 and 1971. Table 5.4 illustrates that over half of the offshore funds launched in this period were for investment in UK equities, reiterating how tax drove the process.

Table 5.4   Types of open-ended offshore funds 1965–71

Year	Totals UK equities	US funds	Rest of world	Total (all)
1965	2			2
1966	3			3
1967	3			3
1968	4			4
1969	4	1	3	8
1970	6	1	4	11
1971	7	1	4	12

Source: Johns (1983, p. 113); and data from Commercial Relations Office
Note: This table includes only open-ended offshore funds so is not exactly comparable with Table 5.3 as it excludes the three closed-ended investment trusts.

Having examined the key factors we next need to consider the supplementary factors: secrecy, customer preference, the 'pin-stripe infrastructure' and tourism.

Secrecy

OFC secrecy stemmed from two sources: traditional banking practice and the legal separation of Jersey from the United Kingdom. First, although Jersey did not have bank secrecy legislation (as Switzerland had), banks operated in the Anglo-Saxon business tradition of client confidentiality. Second, the effective legal separation of the Jersey jurisdiction from the United Kingdom reinforced offshore secrecy and increased the difficulty of information collection by revenue authorities. Again, we return to the underlying relationship with the onshore country.

Many banks, in addition to private banking services, started to offer asset protection such as offshore trusts, often combined with offshore companies. Secrecy was central to this, especially where the assets were being sheltered offshore for tax avoidance (or evasion, which often relies on secrecy from the onshore revenue authorities). Secrecy was also associated with the increasing use of offshore funds, as was noted above, since these provided anonymity for UK and Jersey HNWIs to avoid UK taxes.

Customer preference

In addition to regulation, tax and secrecy, customer preference also affected the banks' move to Jersey. It is already clear that different types of banks were driven by different factors in this period. If we consider the merchant banks – the first banks to arrive in the 1960s – what motivated their colonization of a relatively unknown tax haven island? Merchant banks had specialized in HNWIs for most of their history so that individual customer preference is of great significance. In the 1960s there was increasing customer preference for Jersey over the traditional Caribbean tax havens. As islands in the West Indies experienced increasing nationalism and became independent, customer perception of risk grew so that Jersey seemed to be an alternative, safer haven. This illustrates the combined effect of several factors: Jersey's low taxes, bank secrecy and political stability, which are all based upon Jersey's peculiar status as a British dependency.

Agglomeration

The rapid growth of banking in Jersey was accompanied by an expansion of associated services such as those provided by lawyers, accountants, stockbrokers, fund and trust managers, etc. This 'pin-stripe infrastructure' (Hampton, 1996), which had developed from the increasing demand of wealthy immigrants for sophisticated financial services, was then well placed to offer these services to an international clientele. This agglomeration of financial service firms then attracted further new entrants to the OFC, which in turn expanded its financial infrastructure.

Tourism

The Jersey tourist industry also continued to grow through the 1960s. There is an argument that tourism provided an internal subsidy to the emerging OFC via the shared infrastructure such as airport and communications. However, the OFC activities (particularly the large pool of deposits) generated cheap loans which enabled the financing of new hotels in the 1960s. This interrelation of the two industries suggests more of a two-way flow or symbiotic relationship, rather than a clear case of tourism partly subsidizing the OFC.

Prerequisites for the OFC

Finally, we need to consider location, technology and Jersey's labour force. First, Jersey's location assisted the development of the OFC, being in close proximity to the large financial markets of London and located in the same time zone. Second, rapid improvements in telecommunications allowed the increasing integration of the global financial system, facilitating Jersey-based OFC firms' activities.[13] Thirdly, Jersey had the advantage of an English-speaking and educated labour force with a high literacy rate compared with rival Caribbean OFCs.

By the end of this period, Jersey's financial sector was based on two main strands: wealthy immigrants and financial services. The two were interlinked: as the wealthy settled in the island, so they demanded better financial services, and hence the 'pin-stripe infrastructure' was established and became an important attraction to other OFC firms. For this second phase, regulation and taxation seem to have been working in combination, although arguably the latter was more significant for most new bank entrants. Secrecy was a supplementary factor, although it operated in conjunction with tax, facilitating tax avoidance. All these factors were effectively determined by the specific advantages of the political situation, which hinged on the relationship with the UK mainland.

Conclusion

Jersey's emergence as an OFC clearly shows that although tax was significant, it was not always the dominant factor at all times. At certain times (such as in the late 1960s) regulation was equally important. The analysis of the two phases reveals that regulation and tax were dominant overall, with secrecy and political stability being less significant. Thus far, orthodox international banking theory would probably concur with this analysis. However, here I have argued that a new approach to offshore finance is now necessary, one which views the various 'key factors' in the light of the political economy of the onshore state and its specific relationship to the island. This vital relationship underpins the feasibility of OFCs in small territories.

The continuing existence of the Jersey OFC so close to the mainland economy was possible only through the benevolence of key parts of the UK state (specifically, central government and Bank of England) associated with its historically specific relationship with Jersey. This, in turn, was essentially because of the political economy of the United Kingdom, particularly the dominance of UK financial capital over industrial capital. Financial capital was ascendant by the 1960s, giving it the regulatory and fiscal 'space' for banking innovation, first manifested in the London Eurocurrency market and later in the emergence of small island OFCs such as Jersey.

Acknowledgements

I am grateful to Lino Briguglio, Chris Reid, Tim Rooth and Peter Scott for their helpful criticism of earlier drafts of this chapter. However, the usual disclaimers apply.

Notes

1. Although some authors (e.g. Park and Zwick, 1985) classify London as an offshore finance centre, it is more precisely classified as a global finance centre along with New York and Tokyo at the top of the hierarchy of finance centres (Hamilton, 1986).
2. A distinction can be drawn between major international banks such as Citibank and the small, private 'offshore banks' located in some second- or third-tier OFCs such as Montserrat. These latter banks may be legal entities formed purely to shelter an individual or family's wealth rather than financial intermediaries for public banking activities.
3. Interestingly, at the time of revising this chapter (mid-1996), the island's council is talking about the possibility of the Isle of Wight becoming a tax haven in the future. This would require a major change in the nature of its political relationship with the mainland United Kingdom, and at present the plausibility of this is somewhat unclear.
4. The joint Channel Islands–UK Treasury conference in 1928 on tax avoidance had resulted in the so-called 'bailiff's clause' being inserted into the articles of incorporation of Jersey companies to prevent their being used by UK residents for tax avoidance purposes (*Jersey Evening Post*, 30 December 1936).
5. However, a qualification is given that the data for local residents' deposits may have a margin of error of *c.* 10 per cent, reflecting the difficulty of defining residence owing to the increasing use of intermediaries such as local lawyers.
6. When these funds entered the island, they were mortgaged against property at minimal interest rates using the distinction in Jersey property law between a contract and a judicial mortgage. Therefore, UK funds became transformed into a paper debt instrument, a Jersey contract mortgage (a *hypothèque conventionnelle simple*), becoming more like cash than real estate and thus avoiding UK estate duty (interview with Senator Ralph Vibert, former President of Finance and Economics Committee).
7. UK banking in the 1960s can be divided into two broad sectors: primary banks – the clearers; and secondary banks – all other banks operating in the United Kingdom, including merchant banks, British overseas banks, foreign-owned banks plus the so-called 'fringe' banks (Owens, 1974).

8. Interview with Richard Syvret, Commercial Relations Officer, 13 August 1991, Jersey.
9. This bank was linked to the secondary bank G.T. Whyte, part of the Triumph Investment Trust group (Reid, 1982).
10. For example, the letter of comfort from Hambros was only a few lines long, with no mention of capital adequacy (interview with Derek Short, former Managing Director of Hambros Bank (Jersey) Ltd).
11. Grubel demonstrates how the cost of holding reserves can lower profit margins on commercial lending activities. This may force banks to relocate to centres with less bank regulation and, in particular, lower reserve requirements. This argument partially accounts for the spectacular 1960s growth of Eurocurrencies in London.
12. Again, perhaps this can be viewed as indicative of the fragile nature of the relationship between the Channel Islands and the United Kingdom, showing how easily the dominant partner (the United Kingdom) can unilaterally change the rules.
13. The development of the Jersey money market around 1970–71 illustrates this, where two London money brokers set up Jersey offices to collect funds and then place them in the London markets. This was facilitated by the new technology, which provided an electronic link to the London markets, displaying current price information and currency movements.

References

Aliber, R. (1976) 'Towards a Theory of International Banking.' *Federal Reserve Bank of San Francisco Economic Review*, spring: pp. 5–8.
Bell, G. (1973) *The Eurodollar Market and the International Financial System*. London: Macmillan.
Buckley, P. and Casson, M. (1976) *The Future of Multinational Enterprise*. London: Macmillan.
Cain, P. and Hopkins, A. (1987) 'Gentlemanly Capitalism and British Expansion Overseas, 2: New Imperialism, 1850–1945.' *Economic History Review*, vol. 40: pp. 1–26.
Cho, K. (1985) *Multinational Banks: Their Identities and Determinants*. Research in Business Economics and Public Policy series, no. 8, Ann Arbor: University of Michigan.
Choi, S., Tschoegl, A. and Yu, P. (1986) 'Banks and the World's Major Financial Centres'. *Weltwirtschaftliches Archiv*, vol. 122, no. 1, 48–64.
Coakley, J. (1992) 'London as an International Finance Centre.' In L. Budd and S. Whimster (eds) *Global Finance and Urban Living*. London: Routledge.
Coakley, J. and Harris, L. (1983) *City of Capital*. Oxford: Blackwell.
Cole, K., Cameron, J. and Edwards, C. (1983) *Why Economists Disagree*. Harlow: Longman.
Diamond, W. and Diamond, D. (1990) *Tax Havens of the World*, vol. 3.
Doggart, C. (1981) *Tax Havens and Their Uses*. London: Economist Intelligence Unit.
Doggart, C. (1993) *Tax Havens and Their Uses*, revised edition. London: Economist Intelligence Unit.
Ehrenfeld, R. (1992) *Evil Money: Encounters along the Money Trail*. New York: HarperCollins.
Giddy, I. (1983) 'The Theory and Industrial Organisation of International Banking.' In R. Hawkins, R. Levich and C. Wihlborg (eds) *The Internationalisation of*

Financial Markets and National Economic Policy. Research in International Business and Finance, vol. 3. London: JAI Press.

Grubel, H. (1977) 'A Theory of Multinational Banking.' *Banco Nazionale del Lavoro Quarterly Review*, vol. 125: pp. 349–363.

Grubel, H. (1982) 'Towards a Theory of Free Economic Zones.' *Weltwirtschaftliches Archiv*, vol. 118: pp. 39–61.

Hamilton, A. (1986) *The Financial Revolution*. Harmondsworth: Penguin.

Hampton, M.P. (1994) 'Treasure Islands or Fool's Gold: Can and Should Small Island Economies Copy Jersey?' *World Development*, vol. 22, no. 2: pp. 237–250.

Hampton, M.P. (1996) *The Offshore Interface: Tax Havens in the Global Economy*. London: Macmillan.

Harris, L. (1985) 'British Capital: Manufacturing, Finance and Multinational Corporations.' in D. Coates, G. Johnston and R. Bush (eds) *A Socialist Anatomy of Britain*. Oxford: Blackwell.

Hilferding, R. (1981) *Finance Capital*. London: Routledge.

Hilton, A. (1987) *City within a State*. London: Tauris.

Hymer, S. (1960) 'The International Operations of National Firms.' Unpublished PhD thesis, Massachusetts Institute of Technology.

Ingham, G. (1984) *Capitalism Divided?* London: Macmillan.

Johns, R. (1983) *Tax Havens and Offshore Finance: A Study in Transnational Economic Development*. London: Pinter.

Johns, R. and Le Marchant, C. (1993) *Finance Centres: British Isle Offshore Development since 1979*. London: Pinter.

Langer, M. (1988) *Practical International Tax Planning*, 3rd edition. New York: Practising Law Institute.

Lenin, V.I. (1965) *Imperialism*. Beijing: Foreign Languages Press.

Lessard, D. and Williamson, J. (1987) *Capital Flight: the Problem and Policy Responses*. Washington, DC: Institute for International Economics.

Mendelsohn, M. (1980) *Money on the Move*. New York: McGraw-Hill.

Myers, M. (1970) *A Financial History of the United States*. New York: Columbia University Press.

Naylor, R. (1987) *Hot Money and the Politics of Debt*. London: Unwin Hyman.

Offshore Finance Magazine (1991) *Jersey: A Special Report*, December: pp. 27–37.

Owens, J. (1974) 'The Growth of the Eurodollar Market.' Bangor Occasional Papers in Economics, no. 4, Bangor: University of Wales.

Park, Y. (1989) 'Recent Functional Changes in International Finance and Their Implications for International Finance Centres.' In Y. Park and M. Essayyad (eds) *International Banking and Financial Centres*. Boston: Kluwer Academic Publishers.

Park, Y. and Zwick, J. (1985) *International Banking in Theory and Practice*. Reading, MA: Addison-Wesley.

Pollard, S. (1969) *The Development of the British Economy 1914–67*, 2nd edition. London: Edward Arnold.

Powell, G.C. (1971) *Economic Survey of Jersey*. St Helier: Bigwoods.

Reed, H. (1981) *The Preeminence of international Finance Centres*. New York: Praeger.

Reid, M. (1982) *The Secondary Banking Crisis 1973–75*. London: Macmillan.

Rutherford, W. (1976) *Jersey*. Newton Abbot: David & Charles.

Sassen, S. (1991) *The Global City*. Princeton, NJ: Princeton University Press.

Spitz, B. (1990) *Tax Havens Encyclopedia*. London: Butterworths.

States of Jersey (1993) *Annual Report to the States by the Chief Adviser*. St Helier: States Printers.

Theobald, T. (1981) 'Offshore Branches and Global Banking – One Bank's View.' *Columbia Journal of World Business*, vol. 16, part 4, winter: pp. 19–20.

102 *Mark P. Hampton*

Thrift, N. (1994) 'On the Social and Cultural Determinants of International Financial Centres: the Case of the City of London.' In S. Corbridge, R. Martin and N. Thrift (eds) *Money, Power and Space*. Oxford: Blackwell.

Walter, I. (1985) *Secret Money: The World of International Financial Secrecy*. London: Allen & Unwin.

Williamson, P. (1984) 'A Bankers' Ramp? Financiers and the British Political Crisis of August 1931.' *English Historical Review*, October: pp. 770–806.

6

Offshore Finance Activities in Vanuatu: an Empirical Study of Determinants and Growth

T.K. Jayaraman

The Republic of Vanuatu, an archipelago of about 80 islands in the South Pacific, is located roughly 2,300 kilometres to the east of Australia (see Figure 6.1). Formerly known as the Anglo-French condominium of the New Hebrides, Vanuatu gained independence in 1980 and has come to be recognized for its wide range of services to offshore investors (Economist Intelligence Unit, 1987). The finance centre, comprising various types of offshore activities and supportive onshore institutions, was first established in 1971. Successive governments after independence continued their thrust on promotion of offshore finance centre (OFC) activities through provision of various incentives and enactment of legislative measures, including the latest International Companies Act, which was enacted in 1992.

Promotion of OFC activities and development of tourism, which have been described as the twin engines of growth for Vanuatu, have led to the creation of a dual economy. As a result, a modernized and sophisticated services sector, which is mainly confined to the capital city of Port Vila on the island of Efate, exists side by side with a subsistence-oriented primary sector, spread over the rest of the country's territory and scattered into various remote islands. The outer islands are engaged mainly in the production of root crops for self-consumption and in the secondary but declining export sector activities, including copra and cocoa, for which prospects are not bright in the near and medium terms. Partly because of this factor and partly because of the growing attractions of city life, there has been an exodus of rural youth to Port Vila in search of opportunities in the urban sector (Cole, 1993)

Tourism activities can be successfully dispersed and promoted in the outer islands by attracting the visitors to the location-specific natural scenic spots,

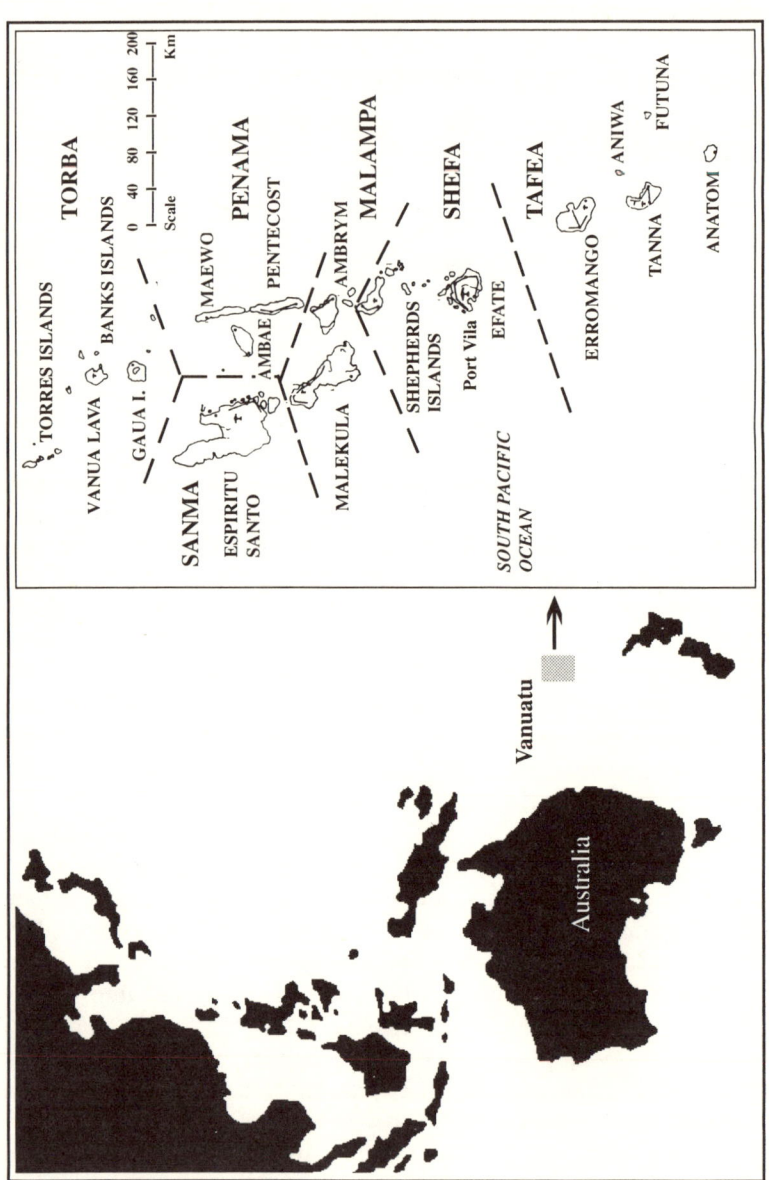

Figure 6.1   The Republic of Vanuatu and its location in the South Pacific Ocean

including live volcanoes, through provision of infrastructures including airstrips and incentives for construction of beach resorts, thereby striving towards balanced regional development to reduce rural–urban disparities (Jayaraman, 1993a). The OFC activities, on the other hand, cannot be so easily spread out as electronic communication facilities, and the pool of specialized services in skills including computer operations, accountancy, law and insurance, which are known as 'pin-strip infrastructures', are firmly urban-based. The question that could be legitimately raised concerns the nature of the impact of OFC on Vanuatu's economic growth and human resource development.

The objective of this chapter is to undertake an empirical study of the factors behind the growth of the OFC in Vanuatu and its contribution to and its impact on the economy. The chapter is organized into four sections: the first section deals with the government measures and growth trends in OFC activities; the second section analyses the factors behind the growth, by testing a simple model developed on the basis of observed practices in Vanuatu and elsewhere in the world, and presents the results of an empirical investigation; the third section evaluates benefits and costs of OFC; and the fourth and final section lists conclusions and recommendations.

Offshore Finance Centre Activities in Vanuatu

Conceptual issues

Offshore finance centres are of two kinds: notional and functional. The notional OFC is a 'paper' concept, which refers simply to loan booking activity, such as the one relating to Eurocurrency centres (McCarthy, 1979). In the 1970s, when Hong Kong residents had to face 15 per cent interest tax, Vanuatu emerged as a satellite centre for Hong Kong as a loan booking centre for the latter's residents desiring to avoid the tax (Johns, 1983) and was a paper OFC before graduating to full status. A functional OFC by definition, on the other hand, offers a full range of financial services, including international banking, offshore fund and trust management, legal services, insurance and a host of related services.

Another related issue is the difference between OFCs and tax havens. Tax havens and OFCs are not necessarily the same thing. Tax havens generally signify either the presence of low tax rates or the absence of any forms of direct taxation. Tax havens so defined, however, may not offer all types of OFC activities. Some tax havens may simply be paper or notional OFCs and continue to remain so without ever reaching the full status.

Tax haven status

Vanuatu has no individual income or corporate taxes and there is a complete absence of any withholding taxes, capital gains or gift taxes, estate duties or death duties for both residents and non-residents. Further, the country has not signed any tax treaties with any country or jurisdiction. Therefore, the country qualifies to be a tax haven in the pure sense of the term. In addition, absence of exchange control with no restrictions on the movement of funds to or from any part of the world has enabled Vanuatu to earn the description 'a happy haven' (Economist Intelligence Unit, 1987).

Vanuatu thus presents 'zero friction' in terms of minimal banking regulations and taxes, and especially absence of direct taxation as against the prevalent 'international friction matrix' on onshore metropolitan countries in the region. The theoretical view is that funds flow from the onshore international friction matrix of bank regulations and taxes that distort the operations of free markets to zero friction (Johns, 1983). This view is confirmed by Walter (1985) and Naylor (1987). They also point out that onshore regulation and taxation would lead to increasing abuse of OFCs, resulting in illegal activities. The positive view, on the other hand, is that the offshore institutions have a proactive role in seeking deposits from the onshore sources.

Macro-economic stability

Vanuatu relies only upon indirect taxes and non-tax sources for raising revenues for meeting public-sector expenditures (Table 6.1). The governments since independence have adopted prudent fiscal and monetary policies and have successfully avoided budget deficits of a sizeable nature on the recurrent budget side (Jayaraman, 1993c). Its development expenditures are mostly financed by external aid, mainly in the form of grants from bilateral and multilateral donors, including the European Union, and loans from the multilateral lending agencies on concessional terms, while technical assistance, all in the form of grants from the donors, bilaterals and multilaterals alike, has been meeting the inadequacies in regard to expertise in various spheres, especially in the public sector. The situation in Vanuatu is not dissimilar to what is generally prevalent in the other island nations in the South Pacific (Jayaraman, 1995b). These policies as well as substantial official development assistance have contributed to relative price and exchange rate stability (Jayaraman, 1993c).

Legislative provisions

Additional facilities, where are generally offered by comparable OFCs elsewhere,[1] are available in Vanuatu. These mainly relate to (a) less complicated regulatory legislation; (b) inexpensive registration and other related operating expenses; (c) secrecy provisions; (d) variety of financial services; and (e) other attractive products. Vanuatu's legal system governing OFC activities consists of the Companies Act 1986 (Chapter 191), under which both local companies and companies exclusively for offshore operations, known as

Table 6.1 Key economic indicators for Vanuatu

Land area (sq. km)	12,000
Sea area (sq. km)	680
Population ('000)	150[a]
Density per sq. km	13
Population growth rate (per cent)	2.9[b]
Gross domestic product (US$ million, 1991 prices)	177
Per capita Income (US$)	1,180
GDP growth rate (%)	2.3[c]
Inflation rate (%)	8.2[d]
Foreign aid (% of GDP)	31.1[e]
Public-sector investment (% of GDP)	16.5[f]
Private-sector investment (% of GDP)	15[f]
Total domestic debt (% of GDP)	31.5[f]
External dent (% of GDP)	21.0[f]
Debt services (% of exports)	2.0[f]
Total government revenue (% of GDP)	23.5[g]
Tax revenue (% of GDP)	18.3[g]
Non-tax revenue (% of GDP)	5.2[g]
Total government expenditure (% of GDP)	40.0[g]
Recurrent expenditure (% of GDP)	21.8[g]
Development expenditure	18.2[g]
Overall budget balance (% of GDP)	-16.5[g]
Recurrent budget balance (% of GDP)	1.7[g]

Sources: World Bank (1991); Asian Development Bank, *Key Indicators of Asian and Pacific Developing countries*, various volumes
[a]1990 estimate
[b]1980–90
[c]1981–90 annual average
[d]1981–91 annual average
[e]Includes all forms of aid including overseas training
[f]1982–91 annual average
[g]1992

exempt companies, have been and can still be registered, and the more recent International Companies Act 1992 is for the specific purpose of facilitating easy registration of OFC companies. In addition, there are other relevant provisions governing OFCs. These are the Banking Act 1970 (Chapter 63), the Trust Companies Act 1971 (Chapter 69), the Insurance Act 1973 (Chapter 82) and the Stamp Duties Act 1970 (Chapter 68).

New International Companies Act 1992

Although any OFC entity, provided that it does not engage in local activities except in furtherance of its international operations, can still be registered under the old Companies Act, the new International Companies Act provides speedy registration procedures, eliminating some of the earlier requirements under the 1986 Companies Act. An exempt company registered under the 1992 Act does not have any annual reporting requirement, unlike in the past. The earlier requirement of one local resident director in Vanuatu has also been done away with. The new Act allows shares to be held in bearer form offshore with only a nominee onshore, with a registered office and a local postal address.

Thus, the new legislation, which enables simple and speedy registration procedures, has considerably reduced the operating costs of an exempt entity registered under the Act. Further, all exempt companies registered under the 1992 Act have been given a guarantee of exemption from taxes of any kind for a 20-year period, except the stipulated annual registration fees. The exempt companies registered under the earlier 1986 Companies Act have been given an option to register themselves under the new legislation. The annual registration fees payable by the exempt entities and the annual fees for work permits for employment/business payable by the expatriate persons are shown in Table 6.2.

Secrecy provisions

The secrecy provisions that existed in the 1986 Companies Act have been further strengthened in the Act of 1992. International investors are immune from the disclosure of any dealings through Vanuatu. The government is also equally committed to its policy of not allowing any information on any transactions or dealings in Vanuatu to be disclosed. Heavy punishments, including fines and imprisonment, are liable to be imposed on those who

Table 6.2   Fees payable by financial sector institutions (in vatu)

Categories	Groups	Annual registration fees	Business Licence Fees
1. Commercial banks	A. Local	200,000	4 per cent turnover; minimum fee: 5 million
	B. Exempt	300,000	—
2. Legal and accounting firms		200,000	4 per cent turnover; minimum fee: 100,000
3. Trust companies and financial institutions	A. Local	200,000	2 per cent turnover; minimum fee: 300,000
	B. Exempt	200,000	—
4. Local insurance companies	A. Local	50,000	A. 1. Life insurance; 2 per cent turnover; minimum fee:150,000 2. Other insurance; 2 per cent turnover; minimum fee: 300,000 3. All classes; 2 per cent turnover; minimum fee: 500,000 B. Brokers
	B. Exempt	100,000	—

infringe the secrecy provisions. However, so far no one has been found guilty of any infringement of secrecy provisions.

Registration of OFC companies

Registration application may be made by nominee companies of the trust accountant, legal and corporate management firms located in Vanuatu. A large number of law partnerships also handle registration and offer legal services. New trust company legislation has been planned in order to strengthen the existing body of law, while retaining the familiar concepts of British trust law and asset protection. A new Company Managers Act is also under preparation so as to give greater regulatory powers to the government to ensure standards of professionalism.

Banking sector

The banking sector, comprising three commercial banks and one fully owned government bank, provides a full range of banking services to enable the exempt companies and investors to move funds into and out of the country. The offshore exempt banks, which are registered and licensed under the Companies Act and the Banking Act, cannot do business in the country, like other exempt companies. However, all these exempt companies and institutions use the banking sector of the country for moving their funds into and out of the country. Absence of exchange control and freedom for both residents and non-residents alike to hold deposits in any currency, either domestic or foreign, have contributed to the emergence of a sophisticated banking system (Jayaraman, 1993b).

Other products

Personal trusts can be set up in Vanuatu against perpetuities. The settlor of a Vanuatu trust can be resident anywhere and does not have to stay in Vanuatu, and the trust property need not be located in Vanuatu either. However, the credibility of Vanuatu trust *vis-à-vis* foreign authorities is enhanced if a local bank or a trust company is appointed as a trustee.

Vanuatu's tax haven status, with its less complicated and less expensive registration procedures, has also encouraged insurance companies to establish themselves as offshore exempt companies. The offshore insurance companies established in Vanuatu undertake captive insurance operations whereby insurance risks in a group are put through the exempt insurance company, which then reinsures through third-party underwriters. The favourable results are reduced premiums and underwriting risks as well as maximum tax deductions across various jurisdictions. The exempted insurance companies so registered in Vanuatu do not have to comply with the financial and reporting requirements which are applicable to insurance companies operating locally.

Ship registration

Another product of Vanuatu's OFC is shipping registration through the Republic of Vanuatu Ship Registry, which offers the full advantages of the International Ship Registry System. It is regulated by legislation introduced in 1981, and the registration service is headed by a commissioner for maritime affairs. A second registration office operates out of New York to make it easier for western-hemisphere shipowners to fly the Vanuatu flag. The initial registration fee for varying tonnages (payable once only), annual tonnage fees and survey tax and other fees charged by Vanuatu are highly competitive and are reported to be much lower than the corresponding fees charged by Panama or Liberia for flying their flags.

Table 6.3 presents data on companies on the live register as of 31 December of each year for the period 1979–93. The data relating to exempt companies are categorized only into three kinds: exempt banks, exempt financial institutions and others. The others include a host of entities including individual trusts and captive insurance companies. Given the limitations in availability of data, it is apparent that the number of total exempt companies of all categories went up from 505 in 1980 to 1,018 in 1992, an increase of 100 per cent. The number, however, decreased in 1993 to 905, possibly reflecting the then prevailing level of investor confidence.

The number of exempt banks increased threefold during the corresponding period, from 41 in 1980 to 120 in 1993. The number of exempt financial institutions, on the other hand, remained steady at 6. The trust companies,

Table 6.3   Companies on the register (as of 31 December each year)

Year	Local companies	Overseas companies		Exempt companies			Total
		Trust companies	Total overseas companies	Exempt banks	Exempt financial institutions	Total exempt companies	
1979	276	12	35	35	5	473	784
1980	291	12	36	41	5	505	832
1981	341	12	34	56	5	555	930
1982	384	13	34	68	5	584	1,002
1983	372	16	36	72	5	516	924
1984	405	15	34	85	7	600	1,039
1985	423	15	40	93	5	644	1,107
1986	403	10	36	93	5	666	1,105
1987	497	12	36	93	6	711	1,244
1988	633	12	37	97	6	736	1,406
1989	812	12	33	103	2	834	1,679
1990	871	12	37	103	6	932	1,840
1991	926	12	37	105	6	890	1,853
1992	1,036	12	40	114	6	1018	2,094
1993	1,123	12	44	120	6	905	2,072

Source: Reserve Bank of Vanuatu: *Quarterly Economic Review*, various issues

which assist the exempt companies in their formation and incorporation, by extending advisory services in legal and tax matters, fluctuated in number from 12 to 16 in the initial years of the period under consideration and in the later years remained steady at 12.

Determinants of Growth of the Offshore Finance Centre

Supply factors

Studies of OFCs have identified certain factors which appear to have contributed to the success of well-known OFCs in the world, such as the Bahamas and Channel Islands. There are (a) tax haven status; (b) commitment to secrecy; (c) an open economy; (d) political stability; (e) a well-developed communications system; and (f) advantages of location and convenience of time zone (International Monetary Fund, 1982). While such a broad generalization may be helpful, certain location-specific characteristics that obtained in the case of Vanuatu, and that cannot be easily replicated elsewhere, deserve mention. These can be broadly categorized into internal and external factors. Internal factors include historic specificity, government initiatives, tourism development, financial infrastructure and the local population's knowledge of English and French.

Historic specificity

Although Vanuatu shared its colonial heritage with other South Pacific island nations before their independence, direct taxation was never introduced during the colonial days. The same policy continued after independence in 1980. Additionally, OFC activities were encouraged by the colonial rulers themselves as early as 1971. This encouragement to Vanuatu to emerge as an OFC is similar to help given by Britain and the Netherlands in regard to OFCs of various Caribbean islands, including the British Virgin Islands, the Cayman Islands, and Montserrat and the Turks and Caicos Islands.[2]

Government initiatives

Government initiatives have been consistently in the direction of moving forward with OFCs. The Minister of Finance made a forthright announcement in the Parliament in December 1993 ruling out any idea of consideration of any form of direct taxation. Further, as noted before, a guarantee of exemption for 20 years has been provided under the International Companies Act of 1992 for all OFC entities registered under the Act. In addition to being a liberal, open economy, Vanuatu has no exchange controls and there are strict secrecy provisions, ensuring confidentiality of all operations by exempt companies. These measures and the planned legislative steps have made clear the government's intention to portray Vanuatu as a place of 'zero friction' in the international friction matrix for attracting financial capital.

Tourism development

The government has been promoting tourism as one of the two engines of growth, the other being OFC. Tourism promotion activities have contributed to the creation of infrastructure in terms of airport development, hotels, resort facilities and many leisure activities, including golf courses and casinos, besides providing support to the national airline so that it can bring in more tourists. All these efforts have popularized Vanuatu among retirees and others seeking places to spend their leisure time, thus attracting tourists and investors alike.

Tourists who visit by air or cruise ships on short day trips learn about OFC activities and related advantages and are encouraged to use the facilities. Retirees, especially high net worth individuals (HNWIs), who visit Vanuatu initially as tourists are encouraged to put their savings in Vanuatu trusts and derive tax advantages. Similarly, pure investors who come for business are attracted by tourist developments and may decide to spend an extended holiday, and even come later as pure tourists. Thus, tourism and OFC seem to be supporting each other. A 'symbiotic relationship suggested by an interrelation of the two industries', as is witnessed in the Caribbean and Channel Islands (Casanegra de Jantscher 1976; Hampton, 1994b), appears to be operative in Vanuatu.

Financial structure

An open economy with absence of exchange control on the movement of capital into and from the country has contributed to the emergence of a sophisticated financial structure. The HNWIs, besides OFCs, have provided a booming demand for private banking and other services.

Knowledge of English and French

Although university and high school graduates as well as technically qualified persons are in short supply, the relatively high literacy rate has provided the country with a labour force able to speak either English or French. Long traditions of an open-door policy to foreigners who wish to work and reside in Vanuatu have contributed to the emergence of a cosmopolitan culture conducive to acceptance and development of OFC activities which are dominated by expatriates. These expatriates supply the critically required skills in computer operations, law and accountancy, which fall under the description of 'pin-stripe infrastructure'.

Demand factors

Demand factors behind the growth of OFC institutions are (a) actions of the metropolitan powers in the region; (b) impact of international events; (c) competition from other OFCs in the region; (d) demand-driven preferences; (e) the time zone; and (f) technological trends.

Actions of metropolitan governments

The history behind the emergence of tax havens shows that it was the existence of certain features in the tax systems of developed countries that allowed taxpayers to take advantage of the benefits offered by tax haven countries. These included more favourable treatment to trusts located abroad than to domestic trusts and the mechanism of tax deferral that allowed taxpayers of high-tax countries to defer income tax payments on income from foreign sources until it was repatriated. These and other escape valves left in their tax systems by high-tax countries encouraged small-resource-based island and other micro-states to create zero friction for attracting funds from these countries (Casanegra de Jantscher, 1976).

However, in recent years, realizing that not all the blame can be placed on tax havens for the losses of tax revenue they suffered, the developed countries have passed several anti-tax avoidance laws. These measures have reduced the tax havens' attractions and made life for tax haven governments not 'all sweetness and light' (*The Economist* Publications, 1990, p. 90). In regard to Vanuatu and other island nations in the south Pacific, Australia and New Zealand have been watchful on tax evaders and have passed far-reaching anti-avoidance legislation (*The Economist* Publications, 1990, p. 98). However, there have been exceptions, as noted in the case of Jersey's growth as an attractive OFC.[3]

Impact of international events

World events of an adverse nature in recent years have apparently helped Vanuatu's OFC institutions to flourish. During the Gulf War, capital flight from politically unstable areas in the Middle East to other, safer havens, including Vanuatu, took place. Similarly, there has been a noticeable capital flight from Hong Kong to many other safer havens as 1997 approached. Since secrecy provisions do not permit any disclosures, it can be surmised that Vanuatu, with its known political and economic stability, will have been one of the main beneficiaries among many safe havens. From the published revenue figures of the government, during the Panama crisis in 1990, the ship registration figures made a quantum jump and revenues soared, as ships preferred to fly the Vanuatu flag (Table 6.4). It is thus obvious that crises elsewhere in the world should result in gains for Vanuatu.

Table 6.4 Shipping registration (as of 31 December each year)

Year	Ships registered	Shipping fees collected (Vt million)
1981	1	n.a.
1982	4	1.0
1983	25	5.0
1984	46	12.0
1985	65	18.5
1986	93	29.1
1987	150	32.3
1988	198	33.1
1989	310	63.0
1990	357	119.0
1991	401	46.2
1992	380	44.0
1993	385	50.0

Source: Reserve Bank of Vanuatu, *Quarterly Economic Review*, various issues.

Competition from other OFCs

Competition from the other South Pacific OFCs including those of Cook Islands, the Solomon Islands, Tonga and Western Samoa has been less than intense for many reasons. Chief among them is that none of the countries can match Vanuatu's total absence of direct taxation. Second, the vigorous campaign and support for OFCs in combination with promotion of tourism have ben a unique feature of Vanuatu, one not paralleled anywhere in the region. Above all, Vanuatu's OFC has so far been free from any damage to its reputation caused by questionable deals of the kind witnessed in regard to the Cook Islands' OFC, as reported in the press (*Islands Business Pacific*, 1994).

Demand-driven preferences

In addition to those investors in Hong Kong and Taiwan who, in the midst of uncertainties, perceive Vanuatu as a safe haven, a large number of retirees and leisure-seeking active business people from other parts of the world, including Australia, find Vanuatu a peaceful and scenic place for spending the rest of their lives and enjoying holidays. As a result, many of them have been moving their funds to Vanuatu.

Time zone

As Vanuatu OFC operates just three hours ahead of the South Asian markets, investors from South-East Asia, especially Taiwan and Hong Kong, find it convenient to deal with Vanuatu.

Technological trends

Improvements introduced in telecommunications, through satellites and electronic mail, have eliminated the need for services to be physically close. Vanuatu's telecommunications system is operated by a joint venture between two well-known multinational private-sector companies and the government. Substantial investment has been undertaken to provide international links on a 24-hour basis using digital technology.

A model for empirical investigation

Although most of the factors described above are of a general, descriptive nature, certain factors can be quantified and testable hypotheses can be formulated. However, construction of a model for undertaking an empirical analysis is severely constrained by the inadequate nature of the data. For example, national income data for Vanuatu are available only from 1983 to 1990, and the data on OFCs, as far as they relate to value-added activities in terms of salaries and wages and profits in non-secret operations, have been compiled purely on a voluntary basis and they cover the period 1981–93.[4] Since the number of observations is constrained by the limited time-series data, the degrees of freedom are insufficient for a robust statistical analysis. In the context of these highly restrictive conditions, an empirical analysis has been attempted with a caveat that the results obtained should be viewed with due caution.

Testable hypotheses

On the basis of the theory of 'zero friction', one can hypothesize that global funds generally flow to the OFC of a country whose revenue-raising efforts, as signified by the ratio of total revenue (taxes and non-taxes, including fees and user charges) to gross domestic product (GDP), are relatively less than those of 'higher-friction states'. In other words, the lower the revenue-mobilizing efforts of a tax haven, the greater would be the number of OFC institutions attracted to the country and their contribution to GDP. However, there is an opposite point of view that OFC activities are less sensitive to local taxes and levies as they generate profits in their own right rather than solely via the avoidance of taxes and levies elsewhere (McCarthy, 1979), and therefore there may not be any significant association between local resource-mobilizing effort in a given tax haven which attracts OFC institutions and their operations. Given this viewpoint, the testable hypothesis is that the contribution of OFC activities scaled by GDP, represented by the term *ofc*, which is the ratio of OFC's contribution to GDP, is negatively related to the ratio of total revenue to GDP, which is represented by the term *tr*.

On the other hand, it is claimed by tax haven countries that the revenues 'hypothetically forgone', through either low tax rates or absence of direct taxation, are more than compensated for by revenues generated through licence and registration fees levied on OFC institutions. The emerging hypothesis from this stand is that *tr* is positively associated with *ofc*. Thus, there appears to be a two-way association in the relationship between *ofc* and *tr*, involving some degree of simultaneity between the two variables, although the nature of direction, either positive or negative, remains to be tested. However, in tax havens with no direct taxation of any kind, whether on their citizens, on corporate bodies, on resident expatriates or on OFC institutions, government revenues are mostly realized from taxes on consumption, with substantial incidence on international trade, since most of the imports are associated with expatriates' consumption. Thus, the ratio of tax revenue to GDP is positively associated with ratio of imports to GDP.

As OFCs' institutions are generally encouraged by tourism development, one can postulate that tourism promotion expenditures by the public sector through campaigns and publicity measures directly influence OFCs' contribution to GDP. It can also be hypothesized that OFC activities are positively associated with political stability, signifying greater flow of funds during periods of peace and order. Similarly, economic stability, represented by a low rate of inflation, and presumably the result of prudent fiscal and monetary policies (Blazic-Metzner and Hughes, 1982), would directly influence the entry of offshore institutions and encourage OFC activities.

A simultaneous equation system model

Based on the above, the proposed model is one of a simultaneous equation system with two endogenous variables (*ofc* and *tr*) and four predetermined variables (ratio of tourism promotion expenditures to GDP, ratio of imports to GDP, rate of inflation as a proxy for economic stability and a dummy variable for political stability) and a lagged variable (*tr* lagged by 1 year).

$$ofc = f(tr, tpe, D, inf) \tag{6.1}$$

$$tr = f(ofc, tr_{-1}, m) \tag{6.2}$$

where *ofc* is the ratio of the OFC's contribution to GDP; *tr* is the ratio of total government revenue to GDP; *tpe* is the ratio of tourism promotion expenditure to GDP; *m* is the ratio of imports to GDP; *inf* is the rate of inflation; and *D* is a dummy variable for political stability, assuming a value of unity during years of stability and zero during years of instability.

Estimated equations

The first equation is exactly identified and the second one is over-identified. Therefore, the estimation had to be performed through a two-stage least squares method. The data employed covering a seven-year period for fitting

Table 6.5   Variables employed in regression analysis

Year	OFC's contribution (% of GDP)	Government total revenue (% of GDP)	Tourism promotion expenditure (% of GDP)	Imports (% of GDP)	Rate of inflation (per cent)
1984	6.23	20.3	0.04	55.20	5.4
1985	8.19	23.7	0.06	58.60	-0.1
1986	11.77	24.4	0.18	50.13	-0.07
1987	7.08	26.1	0.19	56.98	8.8
1988	9.72	26.0	0.31	49.55	10.1
1989	9.64	25.6	0.24	50.02	4.5
1990	9.74	24.1	0.23	62.63	4.3

Source: Statistics Department, Government of Vanuatu, and author's calculations

the equations are given in Table 6.5. Both linear and log-linear formulations were attempted. The log-linear equations emerged as better fits in terms of higher R square duly adjusted for degrees of freedom. The results of log-linear equations are reported here, and are as follows:

$$\ln ofc = 2.703 - 1.005 \ln tr + 0.412 \ln tpe^* + 0.069D^{**} - 0.014 inf^*$$
$$\quad\quad (2.817)\ (-1.533)\quad\quad (5.112)\quad\quad\quad (2.815)\quad (-3.898) \quad\quad\quad\quad (6.3)$$

adj R sq = 0.919   D W Stat = 2.984   F ratio = 18.190

$$\ln tr = -0.377 + 0.048 \ln ofc + 0.724 \ln tr_{-1}^* + 0.422 \ln m^*$$
$$\quad\quad (3.17)\quad (-1.312)\quad\quad (0.607)\quad\quad (6.306) \quad\quad\quad\quad\quad (6.4)$$

adj R sq = 0.921   D W Stat = 2.092   F ratio = 24.247

(The figures in parentheses denote calculated t values.)

* significant at 5 per cent level
** significant at 10 per cent level

In the estimated equation with ln *ofc* as dependent variable, the sign of the estimated coefficient of ln *tr* was negative, indicating a theoretically expected inverse relationship of lower revenue to GDP ratio and higher contribution of OFC to GDP. However, the estimated coefficient was found to be statistically not significant.

The parametric coefficients of other explanatory variables too emerged with the theoretically expected signs. The predetermined explanatory variable *tpe*, representing the ratio of tourism promotion expenditure to GDP, is observed to have a positive association with the dependent variable *ofc*. On the other hand, inflation, which is a proxy variable for economic stability, is inversely related to *ofc*, indicating that a low rate of inflation, representing greater economic stability, attracts OFC activities. These two estimated parametric coefficients were also found to be statistically significant at the 5 per cent level. The dummy variable representing political stability also emerged with statistical significance, although at the 10 per cent level.

In regard to the equation with ln *tr* as dependent variable, the sign of the estimated coefficient of ln *ofc* was found to be in accordance with theoretical expectations. However, the estimated coefficient was found to be statistically not significant. On the other hand, the estimated coefficients of the two predetermined variables, namely lagged ln *tr* and ln *m*, turned out to have the appropriate signs and were found statistically significant at the 5 per cent level.

A single equation model

Since the two-way relationship between *ofc* and *tr* was found to be statistically not significant, a single-equation model with *ofc* as dependent variable was considered more appropriate. Accordingly, the regression equation was estimated by an ordinary least squares (OLS) method, retaining all the four explanatory variables, including *tr*. Again the log-linear equation emerged as a better fit in terms of higher *R* square, duly adjusted for degrees of freedom.

The estimated equation is:

$$\ln ofc = 2.164 - 0.636 \ln tr + 0.376 \ln tpe^* + 0.064D^* - 0.015inf^*$$
$$\quad (2.870)(-1.235) \qquad (5.643) \qquad\quad (2.956) \quad (-4.529) \qquad\qquad (6.5)$$

adj *R* sq = 0.936   D W Stat = 2.499   F ratio = 22.985

As observed before, despite its theoretically expected negative sign, the estimated coefficient of *tr* was found to be statistically not significant. The statistical non-significance of relationship between *ofc* and *tr* indicates that the view taken by McCarthy (1979) that OFC activities are not sensitive to local taxes and levies as they generate profits in their own right cannot be rejected. Perhaps for this reason, it appears that some of the tax havens have been able to impose profit taxes on offshore banks.

Since the variable *tr* was found not to be significantly associated with *ofc*, it was dropped from the estimation procedure and a new equation, retaining all the other independent variables, was fitted by the OLS method. The fitted equation is:

$$\ln ofc = 1.235 + 0.314 \ln tpe^* + 0.054D^* - 0.016inf^* \ (21.189)$$
$$\quad (2.481) \quad (6.592) \qquad\quad (3.026) \quad (-4.478) \qquad\qquad (6.6)$$

Adj *R* sq = 0.925   D W Stat = 2.071   F ratio = 25.636

The estimated equation was found very satisfactory in terms of both high adjusted *R* square and *F* ratio. Further, all the three independent explanatory variables, namely the ratio of tourism promotion expenditure, rate of infla-tion proxy for macro-economic stability (the lower the rate, higher the degree of economic stability) and the dummy variable representing political stabil-ity), not only emerged with the theoretically expected signs but were also found to be statistically significant at the 5 per cent level.

Interpretation of results

Empirical investigation reveals the following: (a) there is no significant association between the contribution of OFC activities to GDP and low-revenue efforts, ruling out any sensitivity of local taxes and levies on OFC activities; (b) there appears to be a confirmed positive relationship between tourism and contribution of OFC activities to GDP, since tourism promotion efforts directly influence OFC activities; (c) political stability significantly influences OFC activities in a positive manner; and (d) inflation has a significant negative impact on OFC activities' contribution to GDP, which means that macro-economic stability is necessary for attracting OFC insti-tutions.

Impact of OFC Activities on the Economy

The impact of OFC activities on the economy of Vanuatu can be categorized into benefits and costs, each of which will be of two kinds, direct and indirect. Generally, direct benefits render themselves into measurable magnitudes. However, since it is difficult to quantify indirect benefits as well as direct and indirect costs, only a qualitative description is attempted in this chapter.

Direct benefits

The main direct benefits of OFC activities are the local expenditures, com-prising operating or recurring expenditures of salaries, wages, rent and interest payments, taxes and levies, and capital expenditures. Table 6.6 provides details of various categories of expenditures in current prices. The salary and wages component has two categories: (a) salaries paid to expatri-ates employed in the offshore institutions; and (b) wages and salaries paid to the local (known as ni-Vanuatu) citizens employed in these institutions.

Expenditure on salaries paid to expatriates has been around 60 per cent of total salary and wage bill (Table 6.7). More recently this proportion increased to 65.8 per cent in 1992 and 71.8 per cent in 1993. On the other hand, the local employees, who received 37.8 per cent of the total wages and salaries bill in 1985, have been getting less in recent years, although their share was about 45 per cent in 1991. However, then the proportion fell to 34.2 per cent in 1992 and 28.2 per cent of total wages and salaries in 1993. However, it will be of interest to note that the proportion of expatriate employees out of the total number employed by OFC institutions decreased over a nine-year period from 28.5 per cent in 1985 to 15.6 per cent in 1993. It is apparent that the

Table 6.6 Local expenditure of OFC institutions (million Vatu)

1985 / 1986

	Banks	Trust companies	Accounting and legal firms	Offshore companies	Total	Banks	Trust companies	Accounting and legal firms	Offshore companies	Total
	1985					*1986*				
Total local expenditure	508	157	263	165	1,093	478	169	280	196	1,123
1. Taxes and levies	15	18	21	65	119	26	20	26	93	165
(a) Central government fees	12	16	18	65	111	22	18	24	93	157
(b) Import duties	2	1	2	n.a.	5	1	1	1	n.a.	n.a.
(c) Local government fees	1	1	1	n.a.	3	3	1	1	n.a.	n.a.
2. Other recurrent expenditures	465	127	211	100	903	434	139	227	103	903
(a) Total wages and salaries	251	68	119	n.a.	438	269	80	135	n.a.	484
(i) Ni-Vanuatu wages and salaries	11	n.a.	n.a.	n.a.	157	139	34	30	n.a.	203
(b) Rent	11	5	14	n.a.	30	38	4	18	n.a.	60
(c) Interest	1	n.a.	5	n.a.	6	6	n.a.	9	n.a.	15
3. Capital expenditure	28	12	30	n.a.	70	18	10	27	n.a.	55

1987 / 1988

	Banks	Trust companies	Accounting and legal firms	Offshore companies	Total	Banks	Trust companies	Accounting and legal firms	Offshore companies	Total
	1987					*1988*				
Total expenditure	691	209	269	201	1,370	855	198	291	250	1,594
1. Taxes and levies	21	14	31	94	160	39	6	18	115	178
(a) Central government fees	17	8	31	94	150	35	5	18	115	173
(b) Import duties	1	4	n.a.	n.a.	5	2	n.a.	n.a.	n.a.	2
(c) Local government fees	3	2	n.a.	n.a.	5	2	1	n.a.	n.a.	3
2. Other recurrent expenditure	594	170	227	107	1,098	752	174	251	135	1,311
(a) Total wages and salaries	345	89	132	n.a.	566	382	104	154	n.a.	640
(i) Ni-Vanuatu wages and salaries	129	36	35	n.a.	200	182	37	39	n.a.	258
(b) Rent	54	8	18	n.a.	80	53	5	29	n.a.	87
(c) Interest	5	1	9	n.a.	15	n.a.	1	11	n.a.	12
3. Capital expenditure	76	25	11	n.a.	112	64	18	22	n.a.	104

1989

	Banks	Trust companies	Accounting and legal firms	Offshore companies	Total
Total expenditure	1,102	240	303	255	1,900
1. Taxes and levies	20	6	20	133	179
(a) Central government fees	16	6	19	133	174
(b) Import duties	3	n.a.	1	n.a.	4
(c) Local government fees	1	n.a.	n.a.	n.a.	1
2. Other recurrent expenditures	973	227	261	122	1,583
(a) Total wages and salaries	309	110	178	n.a.	597
(i) Ni-Vanuatu wages and salaries	146	55	45[	n.a.	246
(b) Rent	57	7	23	n.a.	87
(c) Interest	n.a.	3	14	n.a.	17
3. Capital expenditure	109	7	22	n.a.	138

1990

	Banks	Trust companies	Accounting and legal firms	Offshore companies	Total
Total expenditure	870	313	383	489	2,055
1. Taxes and levies	38	10	30	325	403
(a) Central government fees	31	10	27	325	393
(b) Import duties	4	n.a.	1	n.a.	5
(c) Local government fees	3	n.a.	2	n.a.	5
2. Other recurrent expenditures	768	240	339	164	1,511
(a) Total wages and salaries	366	92	181	n.a.	639
(i) Ni-Vanuatu wages and salaries	148	40	49	n.a.	237
(b) Rent	73	2	22	n.a.	97
(c) Interest	n.a.	6	14	n.a.	20
3. Capital expenditure	64	63	15	n.a.	142

1991

	Banks	Trust companies	Accounting and legal firms	Offshore companies	Total
Total local expenditure	868	273	409	648	2,198
1. Taxes and levies	52	12	20	559	643
(a) Central government fees	33	10	16	559	618
(b) Import duties	4	0	1	n.a.	5
(c) Local government fees	14	2	3	n.a.	19
2. Other recurrent expenditures	778	242	365	88	1,473
(a) Total wages and salaries	382	111	197	n.a.	690
(i) Ni-Vanuatu wages and salaries	208	44	57	n.a.	309
(b)Rent	95	4	27	n.a.	126
(c) Interest	0	10	17	n.a.	27
Capital expenditure	37	18	23	n.a.	78

1992

	Banks	Trust companies	Accounting and Legal firms	Offshore companies	Total
Total local expenditure	1,644	310	390	761	3,105
1. Taxes and levies	91	11	9	565	676
(a) Central government fees	73	10	11	559	653
(b) Import duties	3	0	0	0	3
(c) Local government fees	15	1	0	0	16
2. Other recurrent expenditures	1,505	276	363	187	2,331
(a) Total wages and salaries	686	122	184	85	1,077
(i) Ni-Vanuatu wages and salaries	194	86	46	43	369
(b)Rent	161	11	28	3	203
(c) Interest	—	12	12	3	27
Capital expenditure	48	23	18	8	97

Table 6.6 *cont'd*

	1993				
	Banks	Trust companies	Accounting and legal firms	Offshore companies	Total
Total local expenditure	1,591	280	329	687	2,887
1. Taxes and levies	82	11	11	467	571
(a) Central government fees	62	11	8	467	548
(b) Import duties	3	0	0	0	3
(c) Local government fees	16	0	3	0	19
2. Other recurrent expenditures	1,462	249	311	211	2,233
(a) Total wages and salaries	713	133	193	92	1,131
(i) Ni-Vanuatu wages and Salaries	207	30	38	44	319
(b) Rent	169	6	31	4	210
(c)	—	8	14	3	25
3. Capital expenditure	47	20	7	9	83

Source: Reserve Bank of Vanuatu, *Quarterly Economic Review*, various issues
Note: n.a. = not available

Table 6.7 Employment statistics for OFC institutions

Year	Employment (number)			Wages and salaries (million vatu)		
	Expatriates	Ni–Vanuatu citizens	Total	Expatriates	Ni–Vanuatu citizens	Total
1985	104	261	365	258	157	415
	(28.5)	(71.5)		(62.2)	(37.8)	
1986	98	272	370	281	203	484
	(26.5)	(73.5)		(58.1)	(41.9)	
1987	89	277	366	366	200	566
	(24.3)	(75.7)		(64.7)	(35.3)	
1988	82	307	389	382	258	640
	(21.1)	(78.9)		(59.7)	(40.3)	
1989	75	323	398	352	246	598
	(18.8)	(81.2)		(58.9)	(41.1)	
1990	80	313	393	402	237	639
	(20.4)	(79.6)		(62.9)	(37.1)	
1991	86	311	397	381	309	690
	(21.7)	(78.3)		(55.2)	(44.8)	
1992	70	362	432	709	368	1,077
	(16.2)	83.8)		(65.8)	(34.2)	
1992	68	368	436	812	319	1,131
	(15.6)	84.4)		(71.8)	(28.2)	

Source: Author's calculations; Reserve Bank of Vanuatu, *Quarterly Economic Review*, various issues
Note: Figures in parentheses denote % of total

remuneration package per expatriate employee improved over the period. Correspondingly, the proportion of ni-Vanuatu employees went up from 71.5 per cent in 1985 to 84.4 per cent in 1993.

Although the proportion of salaries paid to expatriates to the total expenditure is sizeable and is classed as local expenditures, it is likely that major parts of the expatriates' incomes are spent on imported consumer goods, and savings are sent home for meeting their overseas consumption needs, including education of children. As a result, net benefits to Vanuatu's local economy from the resident expatriate salaries in terms of backward and forward linkages are likely to be less than would otherwise be the case.

In terms of proportion of GDP, total wages and salaries received by employees in OFC institutions rose from 3.3 per cent in 1985 to 5.1 per cent in 1992 (Table 6.8). Wages and salaries received by ni-Vanuatu employees varied between 1.3 per cent of GDP in 1985 and 1.7 per cent of GDP in 1992.

Taxes and levies

Total recurrent expenditures, besides salaries and wages, include taxes on imports and levies, including registration fees and business licence charges and stamp duties paid to the government and local government councils. The amount paid by OFC institutions in terms of taxes and levies in current prices

Table 6.8   Local expenditure of OFC institutions (% of GDP)

Year	Recurrent expenditure				Capital expenditure	Total local expenditure
	Total wages and salaries	Ni-Vanuatu salaries and wages	Taxes and levies	Total recurrent expenditure		
1985	3.3	1.3	0.9	8.2	0.5	8.7
1986	4.0	1.7	1.3	8.8	0.4	9.2
1987	4.2	1.5	1.2	9.4	0.8	10.2
1988	4.3	1.7	1.2	9.9	0.7	10.6
1989	3.6	1.5	1.1	10.8	0.8	11.6
1990	3.6	1.3	2.2	10.6	0.9	11.5
1991[a]	3.5	1.6	3.3	10.7	0.4	11.1
1992[b]	5.1	1.7	3.2	9.5	0.4	9.9

[a]Estimates based on provisional GDP figures

increased nearly fivefold from Vt 119 million in 1985 to Vt 571 million in 1993. In constant 1990 prices, it went up from Vt 172 million in 1985 to VT 488 million in 1993 (Table 6.9). As a percentage of GDP, revenue derived from taxes and levies charged and collected from OFC institutions constituted about 0.9 per cent in 1985 and 3.2 per cent in 1993. In terms of contribution to total revenues collected by government, revenues from taxes and levies on OFC institutions were well above 10 per cent during 1991–93, having risen from a small proportion of 4.0 per cent in 1985 to the highest proportion of 13.6 per cent in 1991 (Table 6.10). There has been a decline since then, as the corresponding proportions were 13.4 per cent of GDP in 1992 and 11.2 per cent of GDP in 1993.

Table 6.9   Local expenditure for OFC institutions, 1984–93 (Vt million, at constant 1990 prices)

Year	Wages, salaries, rent and interest	Taxes and levies	Capital expenditure	Total local expenditure
1984	1,282 (84.3)	158 (10.4)	81 (5.3)	1,521
1985	1,294 (82.6)	172 (11.0)	100 (6.4)	1,567
1986	1,233 (80.4)	225 (14.7)	75 (4.9)	1,533
1987	1,334 (80.1)	194 (11.7)	136 (8.2)	1,665
1988	1,464 (82.3)	198 (11.4)	116 (6.5)	1,778
1989	1,639 (83.3)	185 (9.4)	143 (7.3)	1,967
1990	1,494 (73.5)	399 (19.6)	140 (6.9)	2,033
1991	1,425 (67.6)	597 (28.3)	86 (4.1)	2,108
1992	2,076 (75.1)	603 (21.8)	87 (3.1)	2,765
1993	1,910 (77.3)	488 (19.8)	72 (2.9)	2,470

Note: Figures in parentheses denote percentages of the total

Total local expenditures

Total local expenditures of OFC institutions in current prices increased nearly threefold during the ten years from 1984 to 1993. Inflation had its own

Table 6.10   Government revenues and taxes and levies paid by OFC institutions, 1985–93 (Vt million)

Year	Tax	Non-tax	Total	Taxes and levies paid by OFC institutions	Taxes and levies paid by OFC institutions (% of total revenue)
1985	2,298	672	2,970	119	4.0
1986	2,310	667	2,977	157	5.3
1987	2,875	624	3,499	160	4.6
1988	3,149	757	3,906	178	4.6
1989	3,380	756	4,166	179	4.3
1990	3,991	944	4,935	403	8.2
1991	3,574	1,136	4,710	643	13.6
1992	3,820	1,234	5,034	677	13.4
1993	3,800	1,273	5,073	570	11.2

Source: Reserve Bank of Vanuatu, *Quarterly Economic Review*, various issues

impact, and taking into account its effect, the local expenditures in constant prices show about a 130 per cent rise over the ten-year period (Table 6.9). In terms of percentages of GDP, local expenditures were 8.7 per cent in 1985 and rose to 14.7 per cent of GDP in 1992.

The major components of local expenditures of OFC institutions are wages, salaries, rent and interest payments. It is of interest to note that there was a declining trend in the ratio of this component to total local expenditures. The ratio was 84.3 per cent in 1984 and it was around 73 per cent on average during 1991–93. On the other hand, the ratio of expenditure on taxes and levies to total local expenditures recorded increases over the ten-year period. From about 10.4 per cent of total local expenditures in 1984, it reached a peak proportion of 28.3 per cent in 1991 and decreased to 21.8 per cent in 1992 and 19.8 per cent in 1993.

Comparison with other OFCs

The available data for comparison with other OFCs (the Bahamas, Bahrain, the Cayman Islands, Panama and Singapore) are rather limited and have not been updated, as they relate to the mid-1970s (McCarthy, 1979). It has been noted that the highest prevalent ratio of total local expenditures to GDP among the countries studies was about 8 per cent. This being so, the current percentage level of 14.7 in Vanuatu compares well with that of other OFCs in the world.

Comparison with tourism

Direct benefits from OFC activities can be compared with those derived from tourism in terms of their respective contributions to GDP. Contribution of tourism to GDP varies from year to year depending on the intensity and the number of cyclones during the season, from November to April each year.

Following the most destructive cyclone, Uma, tourism earnings plunged to the lowest level of 11.5 per cent of GDP in 1987. In normal years prior to 1987, the corresponding ratio was well above 15 per cent of GDP. In recent years (1990–92) owing to intensive tourism promotion efforts, it has been about 25 per cent (Jayaraman, 1993a). Thus, OFC activities, whose contribution to GDP has been about 10 per cent, except for a high figure of 12 per cent in 1986, ranks behind tourism.

Indirect benefits

Indirect benefits, which cannot be quantified with accuracy, are nevertheless important. In qualitative terms, their contribution lies in terms of modernization of the economy. They relate to (a) improvements in the local financial system; (b) development of skills; and (c) linkages with other sectors.

Improvements in the local financial system

The presence of offshore institutions contributes to the development of the local financial system through provision of improved banking facilities. As a result of increased banking practices over the ten-year period, current accounts and savings deposits by ni-Vanuatu citizens in domestic currency have been recording annual increases in recent years, encouraging saving habits and contributing to monetization of the economy.

Development of skills

The number of ni-Vanuatu citizens employed in OFC institutions, which was under 320 and stagnant for three years during 1989–91, rose to about 370 in 1992 and 1993 (Table 6.7). Nearly 50 per cent of them are believed to be involved in technical aspects of financial management and the rest in non-technical aspects, including unskilled areas. Since brain drain through emigration to metropolitan countries is not the serious problem generally witnessed in the Polynesian countries in the South Pacific region, the rate of retention of trained people within Vanuatu is expected to be much higher. As long as the trained persons are retained in the very same OFC institutions or, in the event of horizontal transfers from one to another, among various OFC institutions, the degree of specialization becomes more intensified over time. Except for general clerical and computer-related secretarial work, including reception and front-desk services, the specialized skills gained in OFC institutions are not very relevant to manufacturing, business, retail or trading establishments.

For these reasons, the turnover of ni-Vanuatu citizens employed in OFC institutions is likely to be minimal, and most of them would prefer a lifelong career. To that extent, the number of trained people moving to other establishments with their acquired skills and thereby modernizing an increasing number of existing trading and financial establishments or setting up new ones, and the fresh sets of ni-Vanuatu citizens being trained in OFC institutions, are not expected to be sizeable. Despite these limitations, the view expressed in the early years of OFC in Vanuatu that, if there were no training in OFC institutions, 'the increased prospects will be only for cleaners, messenger boys and house servants' (Keating, 1972) appears, in retrospect, to be too pessimistic, and at any rate, in the context of steady growth in recent years, it is no longer valid.

Linkages with the rest of the economy

As noted earlier, the forward and backward linkages of the OFC activities and their local expenditures would be of a substantial nature only when consumption expenditures are on local items. Although most of the expenditures are on imported items, the demand for services which are mainly value-added activities, such as printing, sales and repair services for office equipment, local transport and telecommunications, have created considerable opportunities for employment and income generation. Externalities in consumption by OFCs for these services are believed to have been substantial.

Direct costs of hosting OFC activities

Costs of hosting OFC institutions can be categorized as direct and indirect. Direct costs are (a) supervision and regulation of OFC entities; (b) provision of telecommunication facilities; and (c) education and training.

Supervision and regulation costs

Although establishment of Vanuatu as an OFC dates back to 1971 and implementation of the existing regulations as well as supervision have yielded considerable experience and skills, new legislative measures such as the enactment of the International Companies Act 1992 and pending legislation relating to trust companies and other entities have entailed considerable effort. Presently, legal expertise in almost all areas have been provided by bilateral assistance. In the absence of any data, it would be difficult to determine how much drafting time has been devoted to new legislation. Further, setting up a new Financial Commission and designating the present Registrar of Companies under the Act as the Financial Commissioner with greater regulatory powers and expanding the office would create additional demand for skills. Most of the incremental number of positions may be filled partly by redeployment from elsewhere within the government bureaucracy and partly by expatriate assistance.

As regards actual supervision, secrecy provisions themselves tie the hands of the government since investors look for as little prying as possible by the officials. In the absence of any strict supervision, the slightest whiff of financial scandal is enough to send investors in search of another tax haven. Secrecy and supervision do not go well together (Casanegra de Jantscher, 1976), and reconciling these seemingly conflicting aspects does impose high costs.

Telecommunications facilities

Vanuatu's infrastructure in terms of provision for power and telecommunications facilities has won international acclaim for efficiency. However, generation, operation and distribution costs are considered the highest in the region (World Bank, 1993).

Education and training

In the context of a widespread lack of high school graduates, skilled persons in accountancy, computer operations and secretarial services in Vanuatu, the OFC entities have drawn the best available manpower by paying attractive salaries. As noted earlier, the training in OFC institutions happens to be highly specialized, and hence the horizontal movement of the trained ni-Vanuatu citizens to other occupations has been negligible. Further, the near-stagnation in employment figures in OFC institutions during the late 1980s limited incremental additions to the pool of trained people.

Indirect costs

The indirect costs include (a) the restricted autonomy of monetary authorities; (b) constraints to the formulating of coherent tax policy; (c) supervision on tax evasion; and (d) possibilities of international crime.

Restricted autonomy

Although there are clear legal provisions that offshore exempt banks and other exempt entities are prohibited from conducting onshore activities, these companies are permitted to undertake operations in furtherance of their activities. These include keeping deposits in the local commercial banks. Such deposits may include current accounts as well as time/long-term deposits. As there is no legal requirement on the part of the commercial banks to keep the deposits of exempt institutions distinct from the rest, the liquidity position of the banks and, consequently, the money supply of the country tends to be overestimated. Further, these deposits are subject to withdrawal by the exempt institutions at any time for transfer anywhere, as there are no exchange controls. Since there is strict confidentiality of banking operations in regard to deposits by exempt institutions, as ensured by the secrecy provisions, the central bank's autonomy tends to be impaired to an extent in regard to control over money supply.

Constraints to formulation of coherent tax policy

In the context of a declining trend in the flows of official development assistance to Vanuatu, affecting the fiscal balances, as well as poor prospects in the world market for primary commodity exports, the government has to step up its efforts to mobilize greater domestic resources. This calls for tax reforms but without adding to the already prevailing highly regressive tax structure, which relies heavily upon trade and commodity taxes. The ministry of finance officials fear that any change in policies relating to taxation may destabilize OFC institutions. Consequently, the reluctance to consider necessary changes seriously constrains the formulation of coherent tax policies.

Supervision of tax evasion

Substantial pressures are continuously being put on Vanuatu by the governments of the metropolitan countries in the region to exercise caution in regard to potential tax dodgers. These pressures have imposed heavy costs on Vanuatu's authorities for keeping a watch as well as answering queries from the concerned overseas governments. Additionally, publicity campaigns for a 'clean image' for Vanuatu to counter the 'discrediting efforts' of outsiders have become a necessity requiring time, effort and money.

Watch on international crime

Vanuatu is intensely aware that OFC activities have the dangerous potential to open the doors 'for admitting permanent linkages between the small island economies and the international criminal world' (McKee and Tisdall, 1990). Although there are penal provisions in the legislation, including powers to confiscate money tainted by laundering or drug operations, it requires enormous efforts on the part of the authorities to exercise these powers as well as caution on the part of the banks in accepting deposits.

The OFCs involved in such questionable deals in many Caribbean islands in the mid-1980s had to wait for several years before regaining international confidence. Authorities in Vanuatu have to be constantly conscious of the compelling fact that any such undetected association with suspected international crime might permanently alter the island's future development options, and the costs involved are substantial.

Evaluation

An overall assessment reveals that the importance of OFC activities lies in their modernizing influence on the economy. Although the activities are confined to Port Vila, monetization of the economy and improved banking practices are the benefits derived on a countrywide basis. In measurable terms, OFC activities' contribution to GDP is about 10 per cent. However, this measures only value-added activities undertaken in OFC institutions. There is a possibility of underestimation since there have been observed deficiencies in reporting by OFC institutions and not all of them report to the official agency in charge of compilation of national accounts.

Although indirect benefits at the outset appear to be substantial, there are many leakages, including imports, to reduce the benefits emanating from the local expenditures of OFC. Employment opportunities have been stagnating, and specialized skills are not very relevant to other non-OFC establishments. Government revenue derived from OFC institutions is modest, at 3 per cent of GDP, and is about 11 per cent of total revenue. On the other hand, direct costs and indirect costs have to be recognized. These mainly relate to supervision and regulation, with emphasis on maintaining the integrity of OFC operations and keeping a constant vigil on questionable deals. Any failure on this account would have a lasting adverse effect and the policy-makers have to examine seriously whether OFC activities can be really relied upon as one of the engines of growth.

Conclusions and Recommendations

Vanuatu now has two decades of experience in OFC activities. Its highly specialized institutions have done well and managed to place the country on the world map as one of the few pure tax havens with a high degree of sophistication and untarnished reputation. The contributions of OFC to growth and modernization have been well recognized and the government has high expectations from them.

In the context of declining prospects in the world markets for the traditional exports of copra and other primary products, the island economies have been forced to find new sources of growth. Accordingly, along with tourism Vanuatu has chosen to rely upon exports of OFC services. However, just as tourism receipts are adversely affected by the destructive impacts of cyclones, OFC services' contribution is susceptible to its reputation and integrity, protection of which is a joint responsibility of the government and OFC institutions themselves. Added to these considerations, one has to be aware of the severe constraints that the tax haven status has imposed on the formulation of national fiscal policies, since any change in domestic tax policies has an impact on OFC operations.

Implications for government

The empirical examination of determinants of OFC activities in Vanuatu, although based on limited data, indicates that local taxes and levies in the past had no impact on OFC institutions. In the context of the current 'high international friction matrix' elsewhere in the world, the government should take full cognizance of its current attractive 'zero friction status' to OFC institutions, especially to exempt companies. There is considerable scope for raising the current level of registration fees, which are applicable to exempt companies, and both registration and business licence fees applicable to other OFC institutions. Such additional imposts on OFC activities may substantially alleviate the highly regressive impact of commodity and trade taxes, which has to be borne to a large extent by the disadvantaged and low-income groups of the country in the absence of direct taxation. The exact quantum of increases in fees has to be examined as part of an overall resource mobilization effort in the context of the government's declared policy of no direct taxation of any kind in the short run.

Implications for OFC institutions

The OFC institutions should realize that their contributions should be more visible and credible. First, it should be recognized that the collection and compilation of data on OFCs relating to employment and local expenditures, especially on various components, needs considerable improvement since it has been observed that there have been lapses in the past in notifying the agencies, in the form of either no reporting or incorrect reporting. The Vanuatu Financial Center Association (VFCA) should consider it in its own interest to highlight its members' achievements and present accurate data on employment and local expenditures. To this purpose, VFCA might do well to establish a research wing for data collection and compilation. A triennial survey of its constituents' expenditure patterns could also be undertaken by the VFCA so as to determine the multiplier effects of its members' local expenditures. Other areas of research could include employment profiles, training achievements and relevant indicators.

In the light of stagnant employment opportunities and consequent limited possibilities in training in OFC institutions, which are not considered very relevant except to similar financial institutions, OFC institutions' isolated but well protected and continued existence during the past two decades may result in tensions between those engaged in subsistence-oriented activities and purely urban-based enclave-type operations. The institutions' current low contribution to government revenues and identity with urban occupation, unlike countrywide tourism activities, may further compound the problem. Having utilized the hospitable climate of Vanuatu, including the pure tax haven facilities, and having drawn upon the best among the limited manpower for two decades, it would be a fine gesture if OFC institutions considered contributing towards human resource development.

The VFCA might consider earmarking funds from its subscriptions for strengthening the middle- and high-school-level curriculum in mathematics through development of teaching materials and sponsoring teachers from overseas in selected schools in the country. Additionally, it might consider offering full fellowships to aspiring employees in OFC institutions and financing their higher education overseas in accountancy and other areas of relevance. In the longer run, VFCA might also consider setting up and running a business training college in Vanuatu by raising funds not only from its constituents but also from fellow expatriate business enterprises.

Acknowledgements

The author would like to thank Mr Julian Ala, Financial Services Commissioner, Vanuatu Financial Services Commission, Republic of Vanuatu, Port Vila, for his useful comments and suggestions. The views expressed by the author in this paper do not necessarily reflect the views and policies of Asian Development Bank.

Notes

1. These include the Bahamas, Bermuda, the Cayman Islands, the Channel Islands, Hong Kong, Liechtenstein, Luxembourg, the Isle of Man, Panama, Switzerland and the British Virgin Islands. For a detailed presentation of these facilities, see Economist Intelligence Unit (1987).
2. Another view is that such measures by Britain and the Netherlands towards supporting OFC aspirations reflected 'their uncomfortable position in their anachronistic role as colonial masters and their unwillingness to continue subsidizing their remaining possessions' (Peagam, 1989).
3. The British authorities' attitude has been particularly positive towards Jersey's initiatives in regard to OFC growth, 'given the island's marginality to mainstream UK policies'. Since the Channel Islands assisted the UK balance of payments to the extent of £100 million each year, there would seem to be acquiescence to the OFC activities, provided that blatant tax evasion or money-laundering is not seen to be encouraged (Hampton, 1994a).
4. It is understood that the Vanuatu Finance Center Association attempted to undertake a survey of employment and expenditures through a questionnaire circulated to its constituents, but the response was found poor, leading to the abandonment of survey efforts.

References

Blazic-Metzner, B., and Hughes, H. (1982) 'Growth Experiences of Small Economies.' In B. Jalan (ed.) *Problems and Policies in Small Economies*. London: Croom Helm.

Casanegra de Jantscher, M. (1976) 'Tax Havens Explained.' *Finance and Development*, March: pp. 31–34.

Cole, R. (ed.) (1993) *Pacific 2010*. Canberra: Australian National University, Research School of Pacific Studies.

Economist Intelligence Unit (1987) *Tax Havens and their Uses*. London: The Economist.

The Economist Publications (1990) *Tax Havens and their Uses*. London: *The Economist*.

Hampton, M. (1994a) 'Treasure Islands or Fool's Gold: Can and Should Small Island Economies Copy Jersey?' *World Development*, vol. 22, no. 2: pp. 237–250.

Hampton, M. (1994b) 'Towards a Theory of Offshore Finance Centers in Small Islands: the Emergence of Jersey as an OFC.' Paper presented at the Islands of the World IV Conference, Okinawa, June.

International Monetary Fund (1982) *Emerging Financial Centers*. Washington, DC: IMF.

Islands Business Pacific (1994) 'Tax Scam in a Winebox: How Cooks' Offshore Business Was Hit Again,' May, p. 53.

Jayaraman, T.K. (1993a) *Tourism Sector in Vanuatu: an Empirical Investigation.* Canberra: Australian National University, Research School of Pacific Studies.

Jayaraman, T.K. (1993b) *Saving Behavior in Vanuatu: 1983–1990.* Canberra: Australian National University, Research School of Pacific Studies.

Jayaraman, T.K. (1993c) 'Fiscal Deficits and Current Account Imbalances in the South Pacific Countries: A Case Study of Vanuatu.' Occasional Paper 4. Manila: Asian Development Bank.

Jayaraman, T.K. (1995a) 'Role of Public Sector in the South Pacific: Problems and Prospects.' *Bank of Valletta Review*, no. 11, Spring: pp. 37–59.

Jayaraman, T.K. (1995b) 'Official Development Assistance to the South Pacific Island Countries: Objectives, Magnitudes and Determinants.' *Journal of the Pacific Society*, vol. 17, no. 4: pp. 25–36.

Johns, R.A. (1983) *Tax Havens and Offshore Finance: A Study in Transnational Economic Development.* London: Pinter.

Keating, D. (1972) 'The International Finance Center'. *New Hebridean Viewpoint*, November–December: pp. 3–6.

McCarthy, I. (1979) 'Offshore Banking Centers: Benefits and Costs.' *Finance and Development*, vol. 16, no. 4: pp. 45–48.

McKee, D. and Tisdall, C. (1990) *Development Issues in small Island Economies.* New York: Praeger.

Naylor, R.T. (1987) *Hot Money and the Politics of Debt.* London: Unwin Hyman.

Peagam, N. (1989) 'Treasure Islands.' *Euromoney*, Supplement, May: pp. 237–250.

Walter, I. (1985) *Secret Money: The World of International Financial Secrecy.* London: Allen Unwin.

World Bank (1991) *World Bank Atlas 1991.* Washington, DC: World Bank.

World Bank (1993) *Pacific Island Economies: toward Efficient and Sustainable Growth*: vol. 1, *Overview.* Washington, DC: World Bank.

7

Analysing the Emergence of an Offshore Banking Centre:[1] the Case of Bahrain*

Brian Kettell

Offshore Banking Activity in Bahrain

Bahrain is an island centrally located in the Arabian Gulf, within 30 kilometres of the eastern province of the Kingdom of Saudi Arabia, 50 kilometres to the west of the state of Qatar, 430 kilometres south-east of Kuwait and 430 kilometres north of the United Arab Emirates (UAE) (see Figure 7.1). Together with the UAE, Bahrain is the financial centre for the Middle East, and has been so since the 1970s when the oil price rise of 1973–74 stimulated economic growth. One consequence of the significant increase in economic activity was the need for an expanded and more sophisticated banking sector.

More specifically, Bahrain discovered oil in 1932, and was the first Gulf state to produce it. However, the government realized that it would be foolhardy to rely totally on a limited resource and one whose value was so volatile. In the 1960s, it embarked on an extensive programme of economic diversification by converting energy resources into higher-value products.

In addition it introduced measures, discussed below, to stimulate the banking sector. This sector is now the third largest employer in Bahrain and makes up more than 16 per cent of gross national product (GNP).

The offshore financial centre was founded in the wake of a government directive of 1975 aimed at encouraging offshore banking units (OBUs). There are now 47 licensed OBUs in Bahrain. The state's convenient time zone, located between those of Europe and the Middle East, has spawned the OBUs' chief activities of international treasury management and foreign

---

* The views expressed in this article are those of the author only and do not necessarily reflect those of ABC or its management.

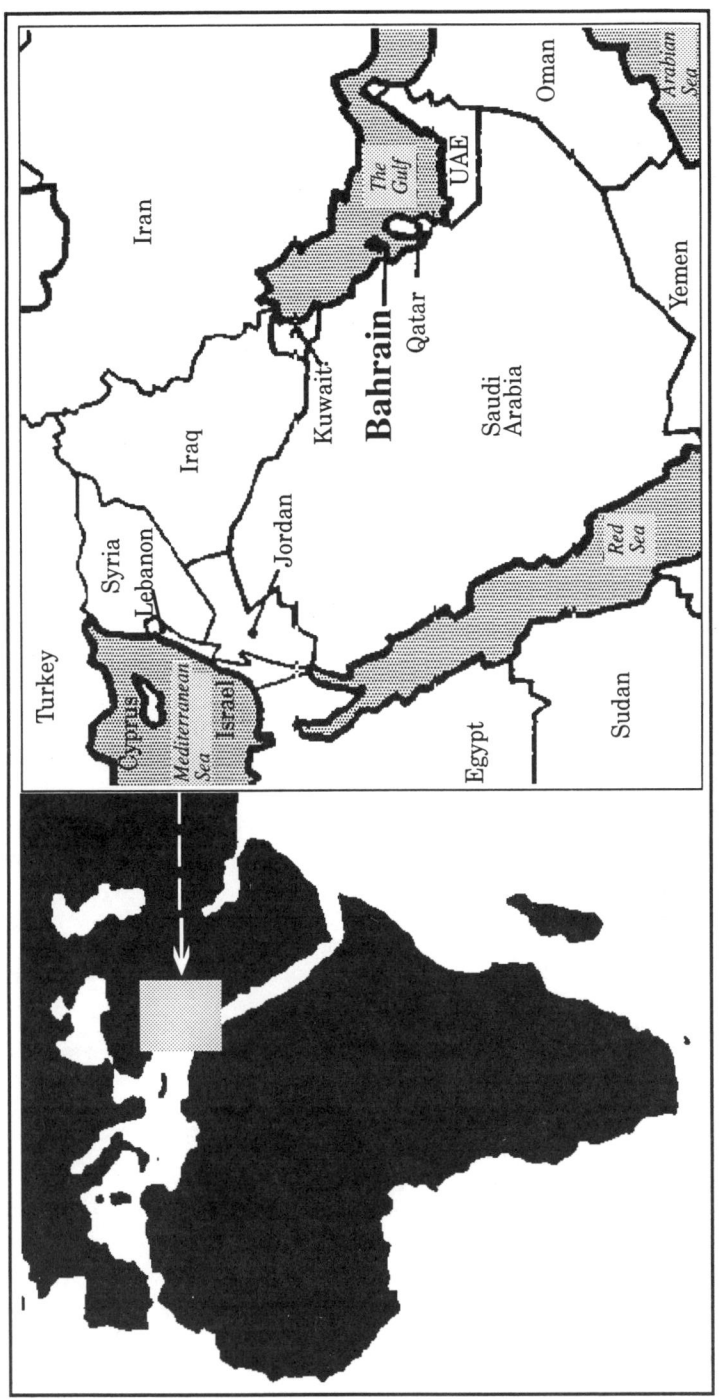

Figure 7.1 The location of Bahrain in the Arabian Gulf

exchange. The civil war in Lebanon and the resulting collapse in its banking industry resulted in many offshore banks relocating to Bahrain.

Barring the UAE, Bahrain tops the list of banking and financial institutions in the Arab world. Out of 1,427 banking and related institutions in 20 Arab countries. UAE has 403 followed by Bahrain (212), Lebanon (144), Egypt (143), Saudi Arabia (117), Oman (86), Jordan (59), Tunisia (48), Qatar (46), Sudan (36), Morocco (30), Yemen (17), Algeria (14), Iraq (11), Libya (11), Djibouti (9), Mauritania (8), Somalia (5) and Syria (5).

Out of the 212 institutions in Bahrain, 47 are OBUs, followed by 17 commercial banks, 4 specialized banks, 22 investment banks, 31 representative banks, 25 money changers, 4 money brokers, 18 insurance companies, 13 accountants and auditors firms, 6 stockbrokers, 9 investment companies and other miscellaneous institutions. The Bahrain Monetary Agency (BMA) has set out the terms and conditions that OBUs must adhere to. These are listed below.

The terms and conditions applying to OBUs

1. OBUs may be branches, subsidiaries or joint ventures.
2. OBUs must be fully staffed and operational at all times.
3. OBUs may transact business with the Government of Bahrain, its agencies, or any licensed bank operating in Bahrain. Transacting business with any other entity or individual resident in Bahrain requires the prior permission of the Bahrain Monetary Agency and such permission would normally be given only if the transaction is related to a development project.
4. OBUs may provide non-residents of Bahrain with all banking services except checking [i.e. current] accounts.
5. OBUs which are branches are not required to maintain any reserves with the Bahrain Monetary Agency.
6. OBUs are required to supply to the Bahrain Monetary Agency such statistical information as may be prescribed by the Agency from time to time.
7. OBUs are required annually to submit to the Bahrain Monetary Agency a balance sheet and profit and loss account audited by auditors approved by the Agency and to submit such other specific or general information as required within 90 days of the year end.
8. OBUs which are partly or wholly owned by banks or other entities are required annually to file with the Bahrain Monetary Agency a copy of the consolidated accounts of the owners.
9. OBUs are required to pay to the Bahrain Monetary Agency an annual licence fee of any amount as may from time to time be prescribed by the Agency.
10. OBUs are not subject to any taxation on profits and no such tax is presently planned or proposed by the Government of Bahrain.
11. OBUs are subject to the provisions of the Bahrain Monetary Agency law – 'Decree law No. (23) or 1973'.

OBU licences have been issued to banks representing interests in the United Kingdom, the United States, Europe, the Far East, South America, Africa and other Middle East countries. However, although there are OBUs representing interests from all parts of the world, there has been a recent move towards

'Arabization'. Major Arab banks like the Arab Banking Corporation (ABC) and Gulf International Bank have established their headquarters there.

As already mentioned, OBUs are allowed to deal in offshore advances and deposits without most of the restrictions and regulations that apply to commercial banks. They can offer all banking services to non-residents, whether government organizations, banks or non-banks. They are not allowed, however, to deal with residents of Bahrain other than government and fully licensed banks. OBUs differ from banks in many offshore jurisdictions in that they are required to be fully staffed, and not just brass-plate operations.

Table 7.1 illustrates the growth of OBUs in Bahrain classified by number of banks and by asset size.

There is a strong element of confidentiality within the banking system in that no one is entitled to have access to any account or any information regarding an account or account holder. The courts, however, may require banks to disclose certain information.

There is no corporate or personal taxation. All profits, dividends or any other income are free from tax. The only exception is for companies engaged in petroleum extraction and refining. Capital, profits, royalties and wages can all be repatriated. Free movement of foreign exchange is also permitted.

Indeed, the only type of tax levied in Bahrain is customs duties. These are generally levied at a rate of 5 per cent for foodstuffs and non-luxuries (electrical appliances and consumer durables), 20 per cent for motor vehicles, 30 per cent for cigarettes and tobacco and 125 per cent for alcoholic drinks.

Table 7.1 Offshore banking units in Bahrain

Year	Number	Total assets (US$ million)
1975	2	1,687
1976	26	6,214
1977	33	15,701
1978	42	23,441
1979	52	27,764
1980	58	37,466
1981	64	50,734
1982	72	59,007
1983	75	62,741
1984	76	62,692
1985	74	56,805
1986	68	55,680
1987	65	63,482
1988	62	68,124
1989	56	72,580
1990	51	59,863
1991	49	53,382
1992	47	69,767
1993	47	60,199

Source: Bahrain Monetary Agency

Activities of Offshore Banking Units

Like all Eurocurrency centres, the Bahrain market is predominantly an inter-bank operation, a regional and international wholesale money centre (see Table 7.2). The type of business is very varied; some OBUs are marketing and administration centres for the area, others provide international wholesale banking services for clients in the Gulf, or are wholesale fund-gathering centres for their head offices. Other OBUs are arms of local banks with regional retail business. Some banks have followed large domestic and multinational customers to the Gulf. Some Arab-owned banks, ABC in particular, use Bahrain as a base for a major international presence. Bahrain has always essentially been a transactions centre rather than a booking centre, and within the overall growth of the market in recent years, Arab banks have played a major role.

Over the period 1980–83, the volume of business of OBUs expanded fast, thereby supporting the thesis that a part of the gap in the international financial markets had been bridged. The general outlook during the period was one of optimism punctuated by caution when the situation so demanded, as happened in 1982 and 1983 when the world recession and the downturn in oil revenues began to bring about a period of consolidation. This had an impact on OBUs' business, and the following three years recorded a steady fall in their total assets/liabilities, as can be seen from Table 7.1, from the peak of US$62.7 billion at the end of 1983 to US$55.7 billion at the end of 1986. This declining trend, the possible impact of the steep fall in oil prices, and the consequent drop in oil revenues spurred the managements of some OBUs to introduce innovative services, to look for new form of business and to compete aggressively to expand their business. The improved perform-ance of industrial countries, faster growth in international trade, higher oil prices and weakening of the US dollar led to the recovery in OBUs' business in 1987, and this continued up to 1989.

During 1990, the time of the Gulf War, total assets/liabilities fell to US$59.9 billion and fell again in 1991 to US$53.3 billion. A swift recovery took place in 1992 to US$69.8 billion. This fell back again in 1993 to US$60.2 billion, a situation which remained the same through the first half of 1994.

At the end of 1993, 58 per cent of OBU funds were inter-bank deposits and 63 per cent of their placements were with banks (see Table 7.2). While inter-bank deposit varied between 63 per cent in 1988 and 58 per cent in 1993, inter-bank loans fell from 78 to 63 per cent during this time period.

The growth in non-bank deposits, rising from 24 per cent in 1984 to 32 per cent in 1993, indicates that OBUs have been successful in securing official and other funds (see Table 7.2). In 1993 OBUs also greatly increased their loans to non-banks, with these having risen from 18 per cent in 1988 to 28 per cent in 1993.

When the market in Bahrain was at an early stage of development,

Table 7.2 Assets and liabilities (%) of offshore banking units

End of period	Assets					Liabilities			
	Loans to non-banks	Inter-bank funds		Other assets	Deposits of non-banks	Inter-bank funds			Other liab.
		Commercial banks in Bahrain	OBUs Outside Bahrain			Commercial banks in Bahrain	OBUs Outside Bahrain		
1984	29	67		4	24	69			7
1985	28	68		4	23	69			8
1986	25	70		5	23	68			9
1987	21	74		5	24	67			9
1988	18	78		5	29	63			8
1989	16	79		5	31	62			7
1990	16	79		5	29	63			8
1991	18	75		7	31	59			10
1992	16	77		7	23	68			9
1993	28	63		9	32	58			10

Source: Bahrain Monetary Agency

government and official institutions in the region relied mainly on European and American financial centres for placement of their surpluses. With the passage of time some funds were being channelled through indigenous banks as well as through branches of foreign banks in Bahrain. However, if banks in Bahrain had been merely a channel for official balances, the market would have been in considerable danger of being too narrowly based and would have been greatly affected in recent years. As it was, the banks had to look for other business opportunities.

Geographically, the sources of OBUs' funds have mainly been Arab countries and Western Europe. However, as can be seen from the liabilities side of Table 7.3, the share of Arab countries' deposits has fallen from 65 per cent in 1984 to 50 per cent in 1993. West European deposits during this time period have remained stable, at around 20 per cent, while North American countries' deposits have risen from 4 per cent to 10 per cent over the same period.

Arab countries, as can be seen from the assets side of Table 7.3, continue to use the largest proportion of OBUs' funds, although this share has fallen from 46 per cent in 1984 to 36 per cent in 1993. The proportion of OBU resources going to North America has filled this gap, rising from 4 per cent in 1984 to 17 per cent in 1993.

Apart from the recycling of surpluses within the region, the OBUs in Bahrain are, therefore, instrumental in providing substantial financial flows to 'other countries' – North American and offshore centres. If Bahrain did not exist as a financial centre these funds would in large measure have gone

Table 7.3 Geographical classification of offshore banking units' assets and liabilities in Bahrain (%)

Year	Arab world	N. America	Assets W. Europe	Offshore	Other	Arab world	N. America	Liabilities W. Europe	Offshore	Othe
1984	46	4	26	10	14	65	4	21	7	3
1985	46	5	23	11	15	67	3	19	7	3
1986	45	6	23	11	14	67	4	19	7	3
1987	40	10	24	10	16	60	6	21	9	4
1988	40	7	21	13	19	62	7	17	8	6
1989	41	10	21	10	18	62	8	18	7	5
1990	42	14	22	8	14	57	14	19	5	5
1991	40	14	23	7	16	60	9	16	8	7
1992	33	20	24	7	16	47	18	20	8	7
1993	36	17	23	8	16	50	16	20	9	5

Source: Bahrain Monetary Agency

direct to other centres. Clearly the proximity of the OBUs plays an important role in attracting funds from Arab countries and the Gulf in particular.

Table 7.4 shows that the US dollar remains the dominant currency, with the share of assets having been 77 per cent in 1984 and 74 per cent in 1993 of assets, although in 1992 it reached 84 per cent. In terms of liabilities the dollar share was 71 per cent in 1984 and 69 per cent in 1993, having recorded a low of 68 per cent in 1985 and a peak of 79 per cent in 1992. The other principal currency group, regional currencies, fell from 15 per cent of assets in 1984 to 9 per cent in 1993 and from 21 per cent of liabilities in 1984 to 10 per cent in 1993.

Table 7.4 Currency breakdown of offshore banking units' assets and liabilities in Bahrain (%)

Year	$	Regional currency	Assets DM	S Fr	Other	$	Regional currency	Liabilities DM	S Fr	Other
1984	77	15	2	3	3	71	21	2	2	3
1985	76	16	2	3	4	68	24	2	3	3
1986	76	14	3	2	4	70	22	3	3	4
1987	79	12	3	2	4	73	17	3	2	4
1988	79	10	3	2	5	74	15	4	2	6
1989	81	10	3	1	5	74	15	4	2	6
1990	76	11	4	1	8	73	14	4	1	7
1991	76	10	4	1	8	71	14	5	1	8
1992	84	6	3	1	5	79	8	5	1	6
1993	74	9	5	1	11	69	10	7	1	11

Source: Bahrain Monetary Agency

Tables 7.5 and 7.6 demonstrate that the maturity structure of both assets and liabilities has remained broadly stable between 1984 and 1993. For assets with a maturity of up to 6 months the figure was 76 per cent in 1984 and

Table 7.5 Maturity structure of liabilities of offshore banking units in Bahrain (%)

Year	< 7 days	8 days–1 mth	1–3 mths	3–6 mths	6–12 mths	12–36 mths	> 3 yrs
1984	18.1	30.5	27.2	14.6	4.3	0.5	4.6
1985	21.1	30.8	25.0	12.6	3.2	0.3	6.8
1986	28.5	30.4	19.4	10.1	3.4	0.9	7.1
1987	23.2	28.2	20.3	14.1	6.2	0.9	6.9
1988	25.4	28.8	21.2	11.3	6.5	0.4	6.2
1989	19.4	32.4	26.3	11.4	4.2	0.6	5.4
1990	23.6	30.4	18.9	12.3	6.8	1.2	6.3
1991	27.3	29.4	17.8	11.2	7.3	0.6	6.2
1992	25.9	31.3	18.4	10.3	9.0	0.6	4.2
1993	22.2	27.6	20.4	16.3	7.3	0.7	5.2

Source: Bahrain Monetary Agency

Table 7.6 Maturity structure of assets of offshore banking units in Bahrain (%)

Year	<7 days	8 days–1 mth	1–3 mths	3–6 mths	6–12 mths	12–36 mths	>3 yrs
1984	17.7	23.0	21.7	14.1	5.2	5.6	12.7
1985	19.3	24.3	22.5	9.9	4.6	6.5	12.6
1986	23.8	22.4	18.8	10.9	4.5	6.2	13.4
1987	24.7	18.2	17.6	14.2	7.7	5.9	11.5
1988	26.6	20.6	19.0	11.1	6.9	4.4	11.2
1989	24.9	21.3	19.9	11.5	7.2	4.2	10.8
1990	29.3	19.0	17.4	9.6	9.6	5.3	9.6
1991	32.5	18.1	16.4	10.3	6.5	5.4	10.6
1992	25.8	21.0	18.1	11.5	8.7	4.4	10.3
1993	20.8	20.6	14.4	15.7	11.4	5.7	11.2

Source: Bahrain Monetary Agency

fluctuated around this figure for the next ten years, reaching 72 per cent in 1993. For liabilities of the same maturity period, the figure was 90 per cent in 1984 and fell slightly to 86 per cent in 1993.

A Case Study: the Arab Banking Corporation

Growth and evolution within and outside Bahrain

The Arab Banking Corporation was incorporated in January 1980 by amiri decree in the State of Bahrain. One of the reasons underlying ABC's formation was the recognition that, despite their established presences in international markets, Arab banks were restricted in the volume of their activities by their comparatively small capital structures. Accordingly, ABC was established with an authorized capital of US$1,000 million, and by April 1981 an amount of US$750 million was fully paid by ABC's original three shareholders: the Ministry of Finance of Kuwait, the Libyan Secretariat of Treasury (later the Central Bank of Libya) and the Abu Dhabi Investment Authority. At the end of 1989 the authorized share capital was increased to

US$1,500 million, and in June 1990 paid-up capital was raised to US$1,000 million through an international share offering. The number of institutional and individual shareholders is now over 2,000.

The growing years

The establishment of ABC in 1980 coincided with a period of surging international liquidity attributable to the major oil price increases of the 1970s and the resulting large current account surpluses of OPEC and especially Arab OPEC members. ABC's initial activities were geared deliberately towards contributing to the deployment of this liquidity through the medium of syndicated international lending. Starting with a capital base that far exceeded that of any other Arab bank, ABC was well placed to lead the Arab world in its bid to capture a position in the international finance community that reflected the growing importance of Arab economies in the global economic structure. ABC quickly forged its mark as a lead manager and arranger of major international loans, and through both its own under-writing power and support from some of its Arab banking associates managed to make up for the lost opportunities in the 1970s, when Arab banks had not been able to handle an appropriate share of the greater and more diverse volume of trade and other business flows between the Arab states and the rest of the world.

After establishing its name in a meaningful way in the syndication mar-kets, ABC settled into a consolidation phase, which would have been necessary even if the market for sovereign lending had not contracted. This period was a 'growth consolidation', during which the bank demonstrated its commitment to developing an international presence and close contact with central banking authorities, correspondent banks and clients world-wide by establishing an extensive network of branches and representative offices in major financial centres. In the late 1980s ABC also initiated a strategy of establishing a direct presence in Arab-world markets.

The decline in sovereign lending due to the sovereign debt crisis of the early 1980s coincided with ABC's development of a broader range of fee-earning services and greater emphasis on exploration of other avenues of lending such as ABC-innovated co-financing in collaboration with the World Bank and project financing. Simultaneously, ABC sought to enhance man-agement, systems and control back-up to increase the efficiency and competitiveness of its operations, reassuring clients and correspondents of its responsibility, reliability and credibility as a business partner. For instance, the development of a computer-based disaster recovery system, which enables all branches and subsidiaries to be monitored centrally in both London and Bahrain, proved its worth during the 1990 Gulf crises and maintained the confidence of the bank's correspondents during a difficult time.

A wide-ranging international network of branches and representative offices

ABC's international outlook was reflected in its early ambitions to establish a wide-ranging branch and representative office network throughout the world's major financial markets. ABC pinpointed London as a logical spring-board, and opened a representative office there in 1981, its first overseas office. In May of the next year, the representative office became ABC's first fully operational overseas branch, functioning as a wholesale commercial bank involved in treasury activities, corporate lending and trade finance. In 1985 ABC's London-based operations moved into new premises just next to the Bank of England. In early 1991 some of the London branch's operations were absorbed into ABC's new subsidiary, ABC International Bank plc (ABCIB; see later section).

ABC's New York branch also opened its door later in 1982 and has gradually expanded its range of banking activities to include not only the financing of Arab-world trade and short-term finance to Latin American countries, but also many aspects of corporate finance. The third branch to open in 1982 was in Singapore, and this branch was expanded into a multi-product corporate banking unit with its activities including a variety of treasury and capital market products, as well as the more conventional direct and syndicated loans and trade finance activities appropriate for this fast-growing region. Inaugurated in 1983, the Milan branch has succeeded in tapping lucrative business from the large trade flows between Italy and the Arab world. The Paris branch was created in 1986 with the upgrading of ABC's previously established representative office. It has been particularly active in trade-related financing between France and the North African Arab countries. In mid-1991 the Paris branch became a branch of ABCIB. Bahrain main branch was constituted in 1986, largely with a view to facilitating the evaluation of Bahrain operations as a profit centre, and to separate head office functions which provided overall co-ordination and supervision over the bank's global network. In 1993 ABC's Tunis representative office was upgraded to offshore branch status.

In addition to the branch network, ABC Group's worldwide activities and client relation were strengthened by the establishment of representative office in Cairo, Casablanca, Hong Kong, Houston, Los Angeles, Rome, Tehran, Tokyo and Tripoli.

ABC acquisitions strategy

ABC's large capital base has always provided the bank with the added option of strategic acquisitions to further business development and balance sheet growth, and in 1982 ABC gained access to the West German merchant banking market by purchasing a 90 per cent stake in the private merchant bank Richard Daus & Co. Bankers (subsequently reorganized and renamed

Arab Banking Corporation – Daus & Co. GmbH, with ABC later increasing its share to 99 per cent). In early 1984 ABC acquired a majority stake in a large Spanish bank, Banco Atlantico, upon its resale to the private sector by the Spanish government, and in early 1985 ABC purchased control of the Hong Kong-based Sun Hung Kai Bank, in 1986 renaming it International Bank of Asia Ltd (IBA). IBA became a wholly owned subsidiary of ABC, and in 1993 ABC privately placed a 20 per cent interest in IBA with the Peoples Republic of China-based China Everbright group and sold a further 25 per cent to the public in Hong Kong in conjunction with the listing of IBA on the Hong Kong exchange. In 1988, through ABC (EC) (see below), the bank acquired a 50 per cent stake in the renamed Banco ABC-Roma SA in Brazil. ABC was also involved in the acquisition of a joint-venture commercial bank in Jordan, Arab Banking Corporation (ABC) Jordan, formerly a securities house which was acquired and restructured as a full commercial bank in 1990. In March 1994 ABC bought 55 per cent of BACMAC Ltd, a UK computer software services company.

ABC and Bahrain

As can be seen from Table 7.7, ABC is an important part of the Bahrain OBUs. ABC assets in 1980 were 5 per cent of total Bahrain OBUs. This figure rose quickly, reaching a peak of 38 per cent in 1991, and later fell back to 31 per

Table 7.7   The growth of the Arab Banking Corporation

Year	Number of Bahrain OBUs	Total assets (US$ million)	ABC assets	Bahrain OBUs - ABC	ABC as % of Bahrain total
1975	2	1,687			
1976	26	6,214			
1977	33	15,701			
1978	42	23,441			
1979	52	27,764			
1980	58	37,466	1,952	35,514	5[a]
1981	64	50,734	4,788	45,946	9
1982	72	59,007	7,892	51,115	13
1983	75	62,741	8,762	53,979	14
1984	76	62,692	11,055	51,637	18
1985	74	56,805	13,066	43,739	23
1986	68	55,680	14,582	41,098	26
1987	65	63,482	17,548	45,934	28
1988	62	68,124	19,127	48,997	28
1989	56	72,580	21,730	50,850	30
1990	51	59,863	20,549	39,314	34
1991	49	53,382	20,451	32,931	38
1992	47	69,767	19,490	50,277	28
1993	47	60,119	18,433	41,686	31

[a]The Arab Banking Corporation (ABC) was incorporated in Bahrain by an amiri decree on 17 January 1980. Operations commenced in the third quarter of 1980. OBU = offshore banking unit.

Table 7.8 Strength, size and soundness of Bahrain banks

Bank	Strength (tier 1 capital)		Size (assets)			Soundness (capital assets ratio)			
	$m	% change	$m	Rank	% change	% latest	% prev.	Rank latest	Rank prev.
1 3 Arab Banking Corporation (12/93)	1,433	1.0	18,433	1	-5.4	7.77	7.28	13	14
2 15 Gulf International Bank (12/93)	521	9.1	7,172	2	13.1	7.26	7.53	14	13
3 23 Investcorp (12/93)	373	15.3	1,339	5	17.4	27.84	28.35	6	6
4 40 United Gulf Bank (12/93)	216	4.6	305	10	2.5	70.18	69.44	1	1
5 41 National Bank of Bahrain (12/93)	211	6.0	1,953	3	13.3	10.80	11.54	12	11
6 44 Bahrain International Bank (12/93)	193	1.1	462	7	37.5	41.72	56.73	2	2
7 45 Bank of Bahrain and Kuwait (12/93)	191	6.6	1,718	4	0.3	11.15	10.48	11	12
8 64 TAIB Bank (12/93)	95	1.7	356	9	8.3	26.71	28.46	7	5
9 66 Bahrain Middle East Bank (12/93)	90	9.3	500	6	17.2	18.04	13.67	9	9
10 68 Faysal Islamic Bank of Bahrain (12/93)	85	18.5	273	12	39.0	31.06	36.44	5	4
11 72 Gulf Riyad Bank (12/93)	71	16.5	223	13	17.0	31.76	22.63	4	7
12 75 Bahraini Saudi Bank (12/93)	62	5.9	294	11	1.0	21.21	20.23	8	8
13 82 Islamic Investment Bank of Bahrain (12/93)	54	1.3	147	14	2.4	36.89	37.29	3	3
14 83 Commercial Bank (12/93)	54	16.1	402	8	18.7	13.34	13.64	10	10

Source: *The Banker*, 1994
Note: The numbers in the first column are respectively the bank's Bahrain ranking and world ranking in December 1993

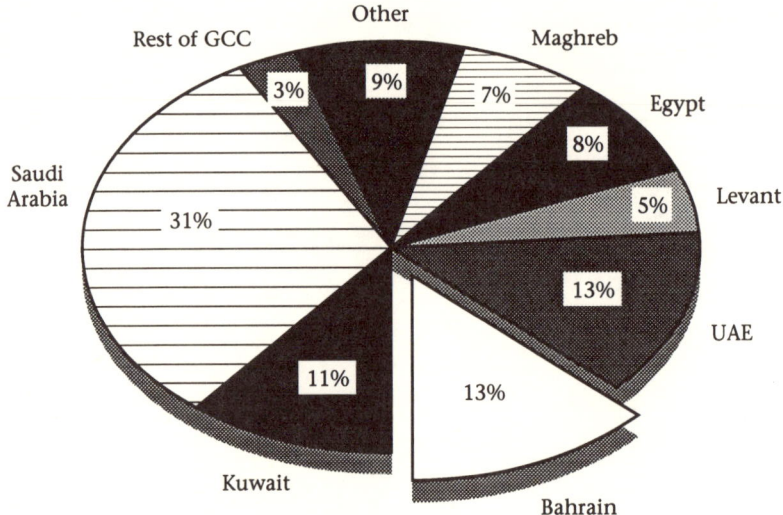

Figure 7.2   Total capital for the Arab 100 = $28.266.6 million ($25,588 million last year's estimate). Source: *The Banker*, December 1994, p. 63

cent. In terms of asset size ABC is the largest Bahrain OBU, as ranked by *The Banker* magazine (see Table 7.8), also the largest Arab bank. However, as can be seen from Figure 7.2, Bahrain represents only 13 per cent of total Arab capital.

A record of strength in an adverse environment

ABC has secured its position in the 1990s as a major banking group which is at the same time Arab in character and international in scope. The ABC Group now employs close to 5,000 people worldwide in its branches, representative offices, subsidiaries and affiliates around the international financial globe.

In order to appreciate fully ABC's performance since its inception in 1989, it must be seen in the context of the evolving international business and economic environment, which at times was far from hospitable to banking and financial activities. The 1980s began with a surge in inflation and interest rates, which caused recession in most of the world's economies. The year 1982 witnessed the onset of the crises in sovereign lending by international banks, the effects of which are still reverberating. Debt rescheduling securitization of existing debt and various products and technological innovations as responses to the problems of sovereign borrowers became a fact of life for the banking community as a result of the sovereign debt crisis. In 1992 and 1993 the crisis in the European Monetary System (EMS) and the resultant turmoil in foreign exchange markets also confronted international banks, while the volatility of bond markets had to be dealt with in 1994.

In the Arab world the 1990 invasion of Kuwait by Iraq provided Arab banks, including ABC, with one of their most severe tests. The ensuing crisis deeply affected the international banking community's confidence in the Gulf region, leading to an abrupt halt in investment in the area. The task of rebuilding confidence in the area, following the cessation of hostilities in March 1991, was made more difficult by the collapse later in 1991 of the Bank of Credit and Commerce International (BCCI) group, which had a negative impact on banks worldwide.

ABC's performance against this backdrop of instability and crises during the 1980s and early 1990s was remarkable, with total asset growth averaging 20 per cent annually, the loan portfolio expanding by 30 per cent per year and total deposits increasing fourfold between 1981 and 1989. The Gulf crisis, however, slowed the pace of growth, as ABC turned to a strategy involving enhanced balance sheet management, deliberately weeding out weaker assets and shifting the group's marketing focus to developing more diverse customer and asset bases, a strategy that has been maintained subsequently. By the end of 1993 total group assets were brought down to US$18.4 billion against US$20.5 billion in 1991 after the crises, while the 1993 loan portfolio of US$9.8 billion was lower than its 1991 level of US$11 billion. During this period total deposits contracted to US$14.3 billion in 1993, although this situation partially reflects a switch to Certificates of Deposit (CDs), bonds and other term financing. As a consequence of the bank's consolidation in the 1990s, ABC achieved 1993 operating profits of US$316 million, the first time it has exceeded US$300 million. Shareholders' funds grew steadily during the 1980s, and at the end of 1990 were US$1,386 million following 1990's share issue. By the end of 1993, after the distribution of profits, shareholders' funds stood at US$1,433 million.

Among the notable features of ABC's growth pattern since its establishment has been its ability to supplement its interest margin income with fee-oriented business following the decline in syndicated lending in the 1980s. Since its inception, the bank has also not lost sight of fundamental banking disciplines. Indeed, it is unique among Arab financial institutions in enjoying a Standard and Poor's credit rating.

Appendix A: SWOT Analysis Applied to Bahrain as an Offshore Banking Centre

Strengths

No taxation

There is no corporate tax, personal income tax or capital gains tax in Bahrain.

History

Bahrain has over 20 years of experience in acting as an offshore banking centre and over 50 years as a trading centre.

Telecommunications

An advanced telecommunications system is already in place. In addition, communication with all the world's major financial centres is possible throughout the working day.

Communications

Good airport services are available. There is some feeling that not enough airlines pass through Bahrain and that this reduces the choices available, but overall the service is believed to be good.

Visas

Visas are easily obtained. A seven-day visa for business visitors can be obtained upon arrival.

Legal and accounting infrastructure

A well-developed legal infrastructure is in place. In addition, there is an adequate supply of international audit and legal firms.

Staff availability

Local staff are available for banking and financial industry positions. It is thought that the availability of local staff is the best in the Gulf. This availability reflects the adequacy of general education. Schools, colleges and the University of Bahrain generate good-quality candidates for trainee positions.

Specialist training

The Bahrain Institute of Banking and Finance (BIBF) provides excellent banking courses. Professional diplomas can be gained and specialist banking courses are taught. The BIBF has an excellent reputation and attracts staff with worldwide reputations.

English-language skills

English is widely spoken. As well as being the medium of instruction at the university, English is heavily emphasized in the schools and, overall, English-language skills are excellent.

BMA support

The Bahrain Monetary Agency (BMA) provides good support and a high commitment to regulation and supervision. This support reflects the overall commitment of the government of Bahrain. During the Iran–Iraq War and after that the Gulf War, the Bahrain government's support to the business community in general and to financial institutions in particular greatly increased international confidence in the country.

Bahrain bankers' club

The presence of the Bankers Society of Bahrain helps bankers to discuss and coordinate banks' concerns with the respective authorities. These concerns normally, however, reflect retail banking rather than wholesale banking concerns.

Social environment

Bahrain provides an excellent social environment for expatriates and their families. They are welcomed by local Bahrainis. There are excellent hotels, restaurants, sports clubs, etc., with the weather to go with them. In addition, there are none of the religious constraints impinging on day-to-day activities that seriously affect other countries in the region.

Licensing

Reputable organizations are speedily dealt with in respect of licence applications. The fact that licences are not given out without appropriate checks is recognized as being essential to maintain confidence in the country.

Weaknesses

Political concerns

Bahrain is located in the Gulf, a politically fragile region. The proximity to Iraq and memories of the Gulf War are still uppermost in the eyes of the financial community. Instabilities in Iran are also cause for concern. These political concerns have resulted in Gulf nationals diversifying their investment away from the region. Over $400 billion in cash from the Gulf Co-operation Council (GCC) countries is estimated, by the Arab Monetary Fund, to have been invested outside the region.

Legal system

Sharia law does not lend itself to commercial disputes of a sophisticated nature. Indeed, the legal system in the Gulf states is not as sophisticated as the banking system.

Another legal consideration is that foreign countries involved in court cases are obliged to submit all documents in Arabic. The translation process is not only costly but also time-consuming. Moreover, a risk in translations of complex financial matters is that many of these are made literally and are consequently sometimes completely out of context.

A recent case in point involved a 'court-appointed expert' handling a court case for an investment bank in which the expert reported contract dealings in 'cucumbers'. The legal dispute centred around losses incurred in commodity futures and options trading. The expert apparently looked up the word 'option' in an English–Arabic dictionary and found the Arab translation to be *Khiar*. But *Khiar* in Arabic can mean either *choice* (i.e. option) or *cucumber*. The so-called 'expert' decided that one cannot possibly trade in 'choices', so the investment instrument must have been a cucumber!

The BMA laws in respect of OBUs' need to be reviewed from time to time. Many BMA recommendations are implemented through circulars rather than laws. This makes the incorporation of internationally recognized community recommendations sometimes slow to be introduced. The BMA has followed BMA guidelines to cover the Basle Committee. GCC countries are included with OECD countries.

Bahrainization

Bahrainization programmes encourage companies to recruit locally. At a senior level this is not thought to be a serious staffing constraint.

Gulf travel restrictions

Travel restrictions for foreigners among Gulf States (e.g. the need to acquire visas) cause lengthy delays. For Bahrain it is thought that there are too many immigration and customs points at the King Fahad causeway which links Bahrain with Saudi Arabia.

Property ownership

There are considerable restrictions on foreign entities' owning of property in Bahrain.

Labour laws

Labour laws do not recognize fully the changing work environment, including redundancies, voluntary termination schemes, staff personnel and housing loans. There is a lack of civil law as a foundation for commercial activities.

Gulf Co-operation Council question marks

GCC countries do not adequately use Bahrain as a financial centre. A distinction is made between a financial centre and a banking centre. If there were more financial transactions passing through Bahrain it would, it is believed, enhance its banking status.

Threats

Fall in petrodollar surpluses

During the late 1970s and early 1980s the Gulf countries were generating significant amounts of petrodollars with very little by way of local banking services to support their activities. Bahrain was the most logical choice to use as a base for Gulf countries to service their needs. The petrodollar surpluses have now largely disappeared. Budget deficits and balance of payments deficits are now widespread in the region.

Competition

Local banking systems, particularly in Saudi Arabia, have rapidly improved and dramatically changed the abilities of Saudi Arabia to recycle the surpluses itself. At the same time other regional countries, particularly Dubai, are fast becoming serious threats to Bahrain.

The global economy

Previously Bahrain filled a convenient gap in the time zone before Singapore/Hong Kong close and European opening. Today, with more advanced communications and information systems, one does not necessarily have to be physically located in a single centre to operate effectively.

Image

Banks in the Arab world are widely perceived to be drab. They are seriously lacking in 'sex appeal'. This is not the case for American or European banks.

Opportunities

Privatization

The GCC privatization agenda should stimulate more innovation within the Bahrain stock exchange. Private-sector participation in state-sponsored activities has occurred since the 1970s but has been limited to specific projects and industries. As the trend towards privatization accelerates, stock exchange turnover will rise accordingly. As liquidity develops GCC funds currently overseas, many then return, with on-going benefits to the banking sector.

Constraints on Saudi banks

The Saudi Arabian Monetary Agency (SAMA) recommends that Saudi banks should adhere to a loans/customer deposit ratio of 65 per cent. As can be seen from Table 7.A1, many banks have exceeded this ratio. Given their inability to lend above this ratio, unless they receive more deposits, opportunities for other lenders within Bahrain may arise.

Table 7.A1   Saudi Arabia: loan/deposit ratios (%) 1990–94[a]

	1990	1991	1992	1993	1994
NCB	35.63	37.80	50.06	53.55	57.10
Riyad Bank	42.62	45.96	50.24	57.32	64.09
Saudi American	34.74	33.78	39.92	43.55	50.11
Saudi French Bank	36.55	43.10	53.82	63.63	66.15
Al Rajhi	115.06	108.96	110.55	113.16	111.57
ANB	35.71	36.96	51.88	60.32	56.75
Saudi Hollandi Bank	38.26	44.67	55.86	65.16	81.11
Saudi British Bank	28.47	35.94	46.02	48.35	48.49
Saudi Cairo Bank	42.98	43.83	47.29	54.65	55.35
Bank al-Jazira	81.97	60.20	67.70	72.69	n.a.
Saudi Investment	23.65	41.20	70.52	67.02	69.22
USCB	28.44	41.27	51.97	67.00	78.62
Total[b]	42.28	45.13	55.16	61.02	66.39

Source: *The NCB Economist*, National Commercial Bank, Jeddah
[a]Up to mid-1994.
[b]Industry total excluding Bank al-Jazira.

Appendix B: The Growth of OBUs in Bahrain: A Statistical Puzzle

Comparing statistics from different sources is always fraught with difficulties. Given different reporting arrangements within the offshore banking world, I compared the official BMA statistics with those supplied by the Bank for International Settlements (BIS) in its publication *International Banking and Financial Market Developments*. As can be seen from Table 7.B1, there is a wide divergence between reported statistics for the size of OBU assets from the two sources. Naturally, I was concerned as to whether this reflected a statistical anomaly or whether there were real differences between the estimates provided by the two sources.

The discrepancy was clarified when further analysis revealed that the BIS data relate to the liabilities and assets of *all* reporting banks in 24 reporting

Table 7.B1 Growth of offshore banking in Bahrain (US$ million): a comparison of BMA and BIS data

Year	BMA[a]	BIS[b]
1985	56,805	13,448
1986	55,680	12,717
1987	63,482	16,518
1988	68,124	16,725
1989	72,580	17,476
1990	59,863	15,182
1991	53,382	14,110
1992	69,767	17,540
1993	60,199	14,995

[a]Bahrain Monetary Agency figures for total assets of OBUs within Bahrain.
[b]Bank for International Settlements figures for external assets of reporting institutions' statistics for Bahrain.

countries *vis-à-vis* entities in Bahrain. There are a number of reasons why the liabilities of the reporting banks (US$17.2 billion in March 1994) *vis-à-vis* Bahrain are much lower than the international assets of the 'offshore banking units' in Bahrain:

- Bahrain assets are those *vis-à-vis* the whole world, while the positions published by the BIS relate only to 24 reporting countries.
- Furthermore, the positions of banks in the United States *vis-à-vis* Bahrain are reported under the residual for the Middle East instead of *vis-à-vis* Bahrain.
- Finally, Bahrain assets include positions *vis-à-vis* non-banks worldwide, while BIS positions are reported by banks only.

Table 7.B2 External positions of banks in individual reporting countries in foreign currencies *vis-à-vis* all sectors

Reporting countries	Amounts outstanding ($US billion)										
	1991 Dec.	1992 March	1992 June	1992 Sept.	1992 Dec.	1993 March	1993 June	1993 Sept.	1993 Dec.	1994 March	1994 June
TOTAL ASSETS	4653.6	4496.7	4595.1	4892.1	4690.7	4690.2	4626.0	4752.6	4725.9	4842.4	4952.2
Other reporting countries	1477.5	1454.3	1438.2	1481.9	1463.5	1470.4	1460.8	1503.7	1469.9	1524.0	1561.7
Bahrain	50.1	54.0	57.0	59.7	66.2	63.0	64.6	57.2	56.4	55.9	60.2
Bahamas	191.1	191.1	167.4	168.0	168.7	170.1	169.5	168.3	167.0	171.3	175.6
Cayman Islands	413.2	413.5	410.8	406.6	404.8	402.8	400.1	397.8	394.0	398.1	401.1
Hong Kong	494.7	483.4	487.2	512.4	496.5	507.2	501.3	538.1	502.9	538.9	550.0
Singapore	322.9	306.2	309.3	327.5	318.9	318.6	316.2	332.9	341.7	351.4	366.3
Other	5.6	6.1	6.3	7.6	8.4	8.7	9.2	9.5	7.8	8.5	8.6

Since I wrote this section, in the November 1994 issue of *International Banking and Financial Market Developments*, the BIS has altered its reporting format. As Table 7.B2 indicates, it has greatly expanded its section on offshore banking centres.

Note

1. I am using the term 'offshore centre' as one in which the international banking sector is subject to more flexible tax or regulatory requirements than the domestic banking market and where intermediation occurs mainly between non-resident lenders and non-resident borrowers.

8

Sovereignty, Security, and the Development of Offshore Financial Centres in the Pacific Islands

Anthony B. van Fossen

A dwarf is as much a man as a giant is; a small republic is no less a sovereign State than the most powerful Kingdom.
> de Vattel, *Le Droit des Gens* (1758, trans. C.G. Fenwick, 1916, ch. 1, s. 18, as quoted in Crawford, 1989, p. 284)

I have long dreamed of buying an island owned by no nation and of putting the World Headquarters of the Dow Company on the truly neutral ground of such an island, beholden to no nation or society.
> Carl Gerstacker, head of Dow Chemicals (as quoted in Sampson, 1988, p. 246)

This chapter explores the dialectic between sovereignty and tax haven development in the Pacific Islands (see Figure 8.1). The most successful tax havens have been internally self-regulating and not subservient to a metropolitan power which is hostile towards haven development. On the other hand, full sovereignty (in the sense of complete political independence) has not necessarily been a crucial advantage and may even have been associated with a lack of proper security (such as that underwritten by a core power) and be an attractor of sleazy operators and clients, as in the case of Tonga. Nevertheless, in the Pacific Islands the offshore financial centres (OFCs) of the tax havens of Vanuatu, Nauru, the Cook Islands and Western Samoa have developed continuously through elaborating legal structures favouring the internationalization of capital. The OFC laws valorize individual appropriation rather than public distribution, minimize state regulation and privilege private ownership. Compared to the second, contrasting group of Pacific Island tax haven (PITH) jurisdictions which we consider next, these states have relatively full sovereignty.

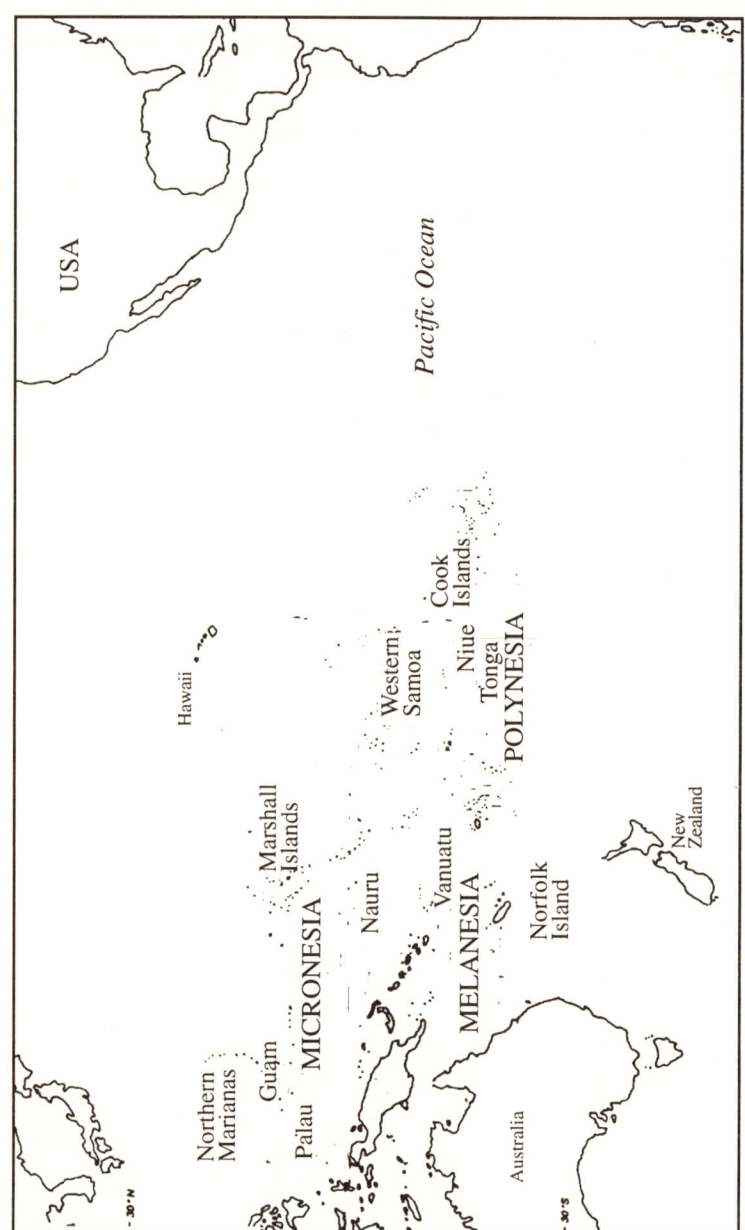

Figure 8.1 Offshore financial centres in the South Pacific

The OFCs of Norfolk Island, the Northern Marianas, the Marshall Islands, Palau and Guam have developed in an uneven manner. They have been constantly influenced by the conflicting prohibitions and requirements of the metropolitan states to which these jurisdictions are closely attached – Australia in the case of Norfolk Island and the United States for the other three. Their OFCs have sometimes been beset by a sense of crisis and ruin largely absent from the OFCs of the first group of sovereign states – since in the second group core powers have had such great influence over how they legally define the rights and character of property. From the mid-1960s until the early 1980s OFC initiatives in these areas were generally opposed by these metropolitan states, but since the mid-1980s, in the American areas at least, the resistance has become more ambiguous and has even turned into qualified support. This relates primarily to the US goals in military strategy and international trade which create particular and not general opportunities for OFC services – in the Marshall Islands, relating to the registration and ownership of flag-of-convenience ships which advance military interests,[1] and in the Northern Marianas and especially Guam, concerning foreign sales corporations which allow American exporters to escape the restrictions of the GATT. Nevertheless, even these new institutional arrangements for the benefit of a particular metropolitan state move towards a more global political economy which is increasingly outside the control of established government regulation. When Norfolk Island became the first OFC in the Pacific Islands in the early 1960s, this constituted an initial and relatively insignificant challenge to a global regime in which a very high proportion of property ownership was defined and regulated by the laws of core states, which postulated and largely enforced an associated set of rights and duties. At present we are in a substantially different world of a vastly expanded number of states, intense ethnic nationalism daily making further claims to sovereignty, and a growing volume of dynamic stateless capital increasingly domiciled in OFCs, which have proliferated around the world and in the Pacific Islands. In the first part of the chapter I will tend to emphasize the advantages of sovereignty for PITHs; in the second half I will qualify this by exploring antithetical tendencies.

OFCs increase the power of transnational corporations (TNCs) over the global system by proposing new stateless financial instruments and arrangements which internalize transactions within the firm and make them less accessible to regulators, taxation officials and competitors. They assist TNCs in maintaining control over valuable knowledge about the world on the most advantageous financial and political terms (as exemplified by Reuters and Dow Jones holding news-gathering, processing and distributing agencies for the Pacific and the Middle East through Cook Islands subsidiaries). They facilitate international mergers which vertically integrate TNCs (as when the diversified oil producer BHP absorbed the giant Hawaiian refiner Pacific Resources through a Cook Islands intermediary). They encourage capital to use cheaper labour in the Third World and threaten protected workers in the core (e.g. by Boral using Vanuatu holding companies and its flag of convenience for new trading vessels). They enhance the bargaining power of TNCs in relation to nation-states by offering capital unparalleled mobility and flexibility (via numerous devices such as the more than 100 offshore banks and 30 captive insurance companies in Vanuatu which allow international financiers to escape exchange control, reserve requirements and other regulations and prudential standards). They encourage Third World elites to place their assets internationally rather than locally – so that they are less likely to be nationalist competitors with TNCs and more inclined to be investors in them (e.g. through Middle East Finance International of Vanuatu, the Primera International Banco of Nauru for Latin Americans, or the elaborate Bank of Bermuda operations in the Cook Islands for Hong Kong Chinese and other Asians). OFCs actively compete with each other, regionally and internationally, to produce laws and services which promote ever greater control of the world by TNCs and the rich.

OFCs are used to minimize the power of governments which attempt to tax 'their' TNCs and rich citizens on their worldwide incomes. The taxation agency *is* the state to a significant degree, and it is strong in the core and weak in the periphery. OFCs have become targets of an international (and particularly an American) offensive against narcotics-trafficking and other criminal activities. This has been used to justify violations of the sovereignty of Third World states (including OFCs) and often places them on the defensive. Worldwide taxation and anti-money-laundering legislation of the United States and some other core countries have produced major contradictions and conflicts in the definition of the sovereignty of OFC jurisdictions in international law.

Yet there is no indication that, even after the collapse of the tax haven-based BCCI, PITH states have lessened their commitment to enacting the most favourable laws for global capital and defending their sovereign right to do so (even in such legally dependent territories as the Commonwealth of the Northern Marianas). Furthermore, intensifying competition between financial institutions encourages their expansion in OFCs, as when the Australian and New Zealand Banking Group entered Western Samoa in 1990 after finding that among its most profitable operations were 'private banking' facilities for rich clients oriented around its branches in the Cook Islands, Vanuatu, Jersey and Geneva. Only the international cooperation of tax offices, policing and regulatory agencies will be effective against OFCs generally. But even anti-OFC countries such as the United States and Australia have done little or nothing to stop the flow of flight capital coming to them from the Third World through OFCs. There are incompatibilities or even contradictions between metropolitan taxation and regulatory systems. And covert action by governments through OFCs leads to frustration of even national efforts against them (as when the CIA allegedly supported the creation of the Cook Islands tax haven, while the Internal Revenue Service naturally opposed it; cf. Wilkes, 1987 a, b, and personal communication, 26 January 1989). Finally, it is unlikely that there will be extensive decriminalization of narcotics-trafficking or other illegal activities which use OFCs, or a substantial decline in the avoidance and evasion of foreign exchange controls and taxation through them.

OFCs make tax avoidance and evasion easier and erode tax morality. They encourage the international mobility of capital and people (as when Hong Kong Chinese use Vanuatu, the Cook Islands and Western Samoa to harbour assets before 1997). They promise a privacy which is not available domestically, and which may be suspended only in relation to some criminal offences such as money-laundering for narcotics-trafficking (e.g. when Vanuatu revealed the financial dealings of Australian marijuana merchants to investigators for Operation Tableau). They depend vitally on a form of trust which is not always honoured and which is not always legally enforceable (as in the numerous bank frauds which have afflicted a number of PITHs and, on a much milder level, the public openness of the Chief Executive Officer of the leading Cook Islands trust company about the affairs of his client, the Bond Corporation).[2] Generally, OFCs encourage short-term planning and liquid (rather than durable) capital formation. They increase the speed of capital to take advantage of opportunities and to avoid taxation, regulation and even discovery.

The size of the annual supply of new stateless capital available to tax havens may be estimated by examining the world's balance of payments deficit with itself. This rough measure of annual additions to global stateless capital was US$41.5 billion in 1987, US$65.5b in 1988, US$100.6b in 1989, US$122.1b in 1990, US$122.4b in both 1991 and 1992, US$88.8b in 1993, US$99.0b in 1994, and US$115.2b in 1995 (International Monetary Fund, 1995; cf. Naylor 1987). These figures are probably understated because of the tendency of governments to want to report current account surpluses. Most of these deficits probably go to tax havens. In addition to this 'official' source of funds, the gross proceeds of international crime have been most attracted to the laundering facilities in OFCs. Estimates of the extent of money-laundering are notoriously unreliable but, according to British intelligence, US$500b may have been laundered globally in 1993 – half from illegal narcotics and the rest from other types of criminal activities and secret organizations oriented towards criminal violence (*The Economist*, 25 June 1994). In addition to this, there is widely believed to be a substantially greater quantity of 'grey' money mediated through OFCs to evade currency control laws, minimize taxes, avoid economic sanctions, finance covert operations and espionage, and provide security for wealthy people in unstable regions, occupations or personal circumstances (*Time Australia*, 18 December 1989). It was estimated in late 1989 that between US$4 trillion and US$12 trillion derived from narcotics-trafficking was deposited in the global banking system (*Offshore Centres Report*, October 1989, November 1990). Banks are the most important intermediaries to tax havens, and the strength of a tax haven is closely related to the depth of involvement by international banks. Rich people and corporations may even establish their own banks in OFCs to strengthen their position in inter-bank transactions, gain anonymity and substantial tax advantages, avoid exchange controls and escape regulations on such matters as maximum interest rates and minimum reserves. But this financial secrecy and lack of accountability produce much higher risks and volatility in the international financial system and help to explain the attempts by metropolitan regulators to impede offshore banking develop-ment (e.g. in Pacific Islands such as Norfolk, Guam, the Northern Marianas and the Marshalls, where core regulators have the greatest power). The increasing number of offshore banks operating in the relatively independent jurisdictions of Vanuatu, the Cook Islands, Nauru and Western Samoa contributes to the continuing growth of their OFCs.

Attempts by metropolitan countries to limit OFCs have generally been haphazard and ineffective – with little or no restriction on the agents or representatives of OFCs operating within their borders or on their own financial institutions operating in OFCs. The interest of financial institutions in OFCs has varied markedly: of the largest four banks in Australia, ANZ and Westpac (each with operations in Vanuatu, the Cook Islands and Western Samoa) have been far more important than National Australia and the Commonwealth. Even in cases involving money-laundering which have been detected, published and prosecuted, the rewards for financial institu-tions operating in OFCs greatly exceed penalties. Particularly since the early 1980s, strict anti-tax haven measures have been approved and computerized surveillance accelerated by most metropolitan countries, but the rates of prosecutions and conviction for tax evasion using OFCs are still extremely low. Many of the same countries opposing tax haven use by their own citizens (such as the United States and Australia) have been creating their own OFCs within their territories to attract foreign stateless capital and give a competitive edge to their own banks, which have also been expanding their OFC networks overseas. It is important to emphasize that OFCs have grown on the basis of the tax assessment ideology of 'residence', which capital-exporting core governments have generally been so adamant in defending and which gives priority to the residence of capital over the place of labour and production. This is significantly related to geopolitical conflict and unequal exchange with the Third World, where tax officials understandably prefer tax assessment by 'source' of income. Metropolitan actions against OFCs are made more difficult since opportunities for companies nominally domiciled in tax havens arise from loopholes and unintended consequences within the core's own predominant tax assessment ideology of 'residence', even when core states assess 'residents' on their worldwide incomes. This is oriented around the larger hegemonic ideology of a world of equal sovereign states in which a person or company can 'reside'.

Taxation and regulation based on 'residence' and a system of equal sovereign states have made extra-territorial approaches and even intergovernmental cooperation between metropolitan states and OFC jurisdictions extremely problematical. Even if mutual legal assistance treaties with tax havens are negotiated, they may have no effect. Some tax treaties with tax havens have resulted in widespread 'treaty shopping', whereby third-country residents use holding companies created in the tax haven (the second country) to avoid paying taxes in their own country or the first country where they have income-producing investments (cf. Gild, 1990) – as expedited in 1991, when leading OFC promoters in Vanuatu took over the largest trust company in Fiji (*Pacific Islands Monthly*, June 1991), which is on the 'whitelist' of virtually every metropolitan tax office and has tax treaties with a number of countries, including Australia. Profits often metamorphose (e.g. from dividends to interest to royalties) as they cross each border. 'Interest' or 'royalty' payments may even be tax-deductible, perhaps in one or more countries – making the problem even more intractable for metropolitan taxation officials.

Frustrations with the difficulties of bilateral and multilateral approaches have led to proposals for extreme action against non-cooperating OFCs. The Gordon Report to the US Internal Revenue Service recommended prohibiting US banks from conducting transactions in such centres, denying banks operating there permission to do business in the United States, applying a withholding tax of 59 per cent on dividends and interest payments to residents of these OFCs, taxing all loans from their institutions to US residents as ordinary income, denying purchasing rights to US assets and withdrawing voting rights from shares held there. There would be a complete suspension of air services between these tax havens and the United States, an end to the tax deductibility of expenses for business done there, and a drastic increase in visa and customs barriers (Gordon, 1981).

However, large and increasing US balance of payments deficits and the need to finance them through foreign borrowing on stateless money markets have made these proposals ever more impractical. Although pressure was exerted against the development of full OFCs in Norfolk Island, Guam, the Northern Marianas and the Marshall Islands (jurisdictions over which metropolitan states have considerable control), even here there was a selective encouragement of some OFC initiatives in the mid- to late 1980s – as when the problems associated with the US balance of payments deficit led to the creation of export-oriented tax haven vehicles named foreign sales corporations (FSCs), primarily in Guam, to avoid residence-based GATT rules about tax concessions to US exporters (Gray, 1985).

Thus metropolitan legislators may intentionally place loopholes in tax legislation which facilitate the use of OFCs. With regard to more sovereign states such as Vanuatu, the Cook Islands, Nauru and Western Samoa, metropolitan states increasingly seek to achieve their ends through cooperation with OFCs, but this occurs rarely and almost always on a case-by-case basis. Agreements between metropolitan states to combat tax evasion through the International Monetary Fund, the OECD and the European Union have been vague and inconsequential – reflecting the sharp conflicts of interests between member states. Even the successes of multilateral anti-money-laundering efforts have been more rhetorical than substantive.

The rhetoric of the metropolitan writers of tax law must be separated from the realities of tax administration to discern the lack of significance of most anti-OFC measures. The rich who use OFCs are the most prepared to pay for secrecy. They use their greater political influence to insert loopholes into the tax laws and associated regulations of their own countries. They have the greatest propensity to hire lawyers, accountants and economists to design avoidance and evasion schemes. If these are discovered or challenged, they are the least likely to be intimidated by the tax office (Peters, 1991). They have the most extensive resources to employ effective lawyers to defend their interests. One may question the ability and even the inclination of the tax office for intense battles with members of the ruling elite (whose financial records may be protected behind the secrecy laws of OFCs). At best we may expect a compromise or a negotiated settlement of cases where tax evasion has been uncovered.

OFC laws are created by lawyers (in collaboration with accountants and increasingly economists). Continual legal innovations serve the interests of the 'yuppies' who are most likely to be at the forefront of tax haven work. Benefits accrue to lawyers according to their ability to write new OFC laws exploiting loopholes in metropolitan legal codes, then to write new metropolitan countermeasures, and finally to start the whole process over again. The esoteric complications of many metropolitan countermeasures against OFCs and the vast amounts of money which depend on their meaning and interpretation create an extremely lucrative specialty – at least in Anglo-American countries, where the legal system serves as both a model and a source of loopholes for OFC laws.

The power of the City of London and the relative autonomy of British lawyers from responsibility to their state and society have given the United Kingdom a particularly powerful role in the formation of OFCs. These OFCs have become a sort of invisible empire of tax havens in present and past British colonies – from the Cayman Islands, the Bahamas and Bermuda to the Channel Islands, the Isle of Man and Cyprus, to Singapore and Hong Kong. The successes of the OFCs of Vanuatu, the Cook Islands, Nauru and Western Samoa are clearly related to the common British basis of their tax haven laws. The problems of the OFCs of Guam, the Northern Marianas and the Marshall Islands are not unconnected with the relative unattractiveness of their American-based legal systems. Liberia and Panama are the only prominent tax havens with laws based on US prototypes. OFC structures usually rely on artificial and formal structures, and the tendency of English jurisprudence towards textualism and literalness assists these constructions, whereas they face far more problems in the American system, with its inclination towards purposive and substantive approaches (cf. Atiyah and Summers, 1987, pp. 100–101).

Even so, the American framework is far superior to most continental European legal systems for fostering OFC development. For example, OFCs are virtually absent in the current and former French empire. There are no OFCs in New Caledonia or French Polynesia, despite the absence of personal or corporate income taxation there. It is difficult to form artificial 'resident' entities within French law, income tax and self-assessment are less important in French public finance, and French lawyers have been restricted in their ability to use legalisms to challenge their state's taxation office. The very different salience of OFCs within the British, American and French spheres of influence also reflects the balance of power between the tax office and finance capitalists.

PITHs may be radically unequal in value because of the unevenness of their legal development. The very small American entrepreneurs developing the OFCs of the Northern Marianas and the Marshall Islands in the early 1980s had little of the professional legal, accounting or economic expertise required to make their tax haven laws attractive to sophisticated clients. The reputations of all these PITHs were damaged through massive fraud. Similar problems occurred in the Tongan tax haven. On the other hand, much larger organizations which are committed to profitable, long-term tax haven development have been important in forming the legal environments of the OFCs of Vanuatu, the Cook Islands, Western Samoa, Guam and (since 1990) the Marshall Islands. This second group is consequently in a much stronger position to attract the lucrative tax haven business of the rich establishment and large corporations through sensitively perceiving and even creating the laws of various jurisdictions. In contrast, the first group of PITHs exploited and defrauded tax evaders who were relatively ignorant of the law. While they exploited naive clients, the second group of PITHs has the legal, economic and accounting resources to translate their powerful and sophisticated clients' objects into laws which strategically serve their interests within national and international legal frameworks.

The great advantage for lawyers and financial institutions of being involved in OFCs is that they attract very rich clients. Furthermore, these rich people are likely to remain highly profitable customers for a long time. They often request discretion and confidentiality about their affairs in OFCs – which suggests the desirability of long-term relationships to avoid misunderstandings and to preserve secrecy. The second group of PITHS are formalizing and legitimating these financial secrecy arrangements in their laws. In this group the most sovereign jurisdictions have the greatest power to do so and the American unincorporated territory of Guam has the fewest possibilities. The most successful PITHs will probably follow the lead of Switzerland in vigorously defending sovereignty and neutrality, the secrecy of financial transactions and an elaborate set of legal institutions which scrupulously favour the interests of corporations and the richest people on earth.

However, there are sometimes problems for PITHs when lower-ranking enforcement officials with significant knowledge of tax avoidance or evasion strategies take the initiative – as when about 37 local Australian Taxation Office (ATO) inspectors conducted a raid on Citibank offices in June 1988 to learn more about an alleged scheme involving large Australian companies sending hundreds of millions of dollars through the Cook Islands (*Citibank* v. *Federal Commissioner of Taxation* (1988) 88 Australian Tax Cases 4,598). This independence of action was widely castigated by conservative commentators, and the ATO hierarchy reasserted its control and supervision over such branch offices. Unusual precautions were invoked by the court to avoid publicity concerning the details of the case. Little was revealed about the activities of banks and companies. The ATO's confrontational strategy has been generally replaced with a more cooperative and conciliatory approach. While organizations such as the ATO have enormous legal powers to enforce tax laws, they are sometimes faced with the ability of politically influential millionaires, high-priced lawyers and unsympathetic judges and politicians to hinder their work.

The courts are often avoided in disputes related to the use of OFCs because they are costly and because they frequently resolve disputes by excessively favouring one of the disputants and disproportionately punishing the loser. Tax offices still use lawyers extensively but they are often reluctant to refer matters to courts. This gives OFC lawyers greater latitude and hope for a negotiated compromise for any irregularities or evasions which the tax office may discover. Tax offices, in their turn, often prefer to use resources for auditing rather than for expensive courtroom confrontations. However, the court is still extremely important for determining many tax principles and enforcing judgments. The court is an ultimate threat to both the users of OFCs and the taxation office. Its ideological orientation greatly affects the probability of success for each party – as when the recent rightward shift in the US Tax Court led to decisions against the Internal Revenue Service whereby companies are increasingly allowed to consider as tax-deductible the premium payments which they make to their own wholly owned captive insurance companies (Winslow, 1990), or when the Barwick High Court in Australia provided perhaps the greatest source of demoralization for the ATO in the early 1980s. Furthermore, the court may set the agenda for further tax legislation – where legislators either extend or negate its judgments.

Few publicized trials involving OFCs concentrate on offences which merely involve the evasion or avoidance of tax. There is usually some powerful non-tax reason why the defendant is unpopular among the powerful and the wider population (as in the trials of executives of Equiticorp, the New Zealand company whose collapse led to hundreds of millions of dollars worth of losses for investors).[3] The coupling of alleged tax evasion and other acts which are labelled antisocial give the taxation office and the court added legitimacy.

Of course, there is struggle over whose law should be dominant: that of the OFC or that of the metropolitan state. This conflict can produce unexpected twists. One of the most significant recent examples is the asset protection trust, developed largely within Cook Islands law. It is designed to place assets outside the surveillance, judgment and regulation of metropolitan courts. Asset protection trusts are intended to transfer the only appropriate and binding areas of litigation and legal judgment from the metropolitan state to the OFC jurisdiction. This attack on metropolitan legal dominance is so bold and new that a few relatively sovereign OFCs such as the Cook Islands have prospered from offering asset protection trusts, while metropolitan courts struggle to find successful legal definitions of them. This may be particularly difficult in metropolitan jurisdictions where law concerning trusts is relatively underdeveloped or even absent. The Cook Islands (like Vanuatu, Nauru and Western Samoa) are attractive for the formation of trusts because they draw on British trust law (generally considered to be superior to that of any other nation) – although the Cook Islands asset protection trust law adds American features which enhance its attractiveness to US residents. Continuing improvements in the Cook Islands asset protection trust laws are likely to make it even more difficult for metropolitan courts to deal with them. Much also depends on conflicting metropolitan judicial interpretations of the fairness of large monetary settlements for professional (including medical) malpractice, alimony and personal and commercial damages, which Cook Islands asset protection trusts are designed to avoid or evade. The threats posed to claimants by these asset protection trusts have already led to quicker, less formal and sharply reduced settlements. The sovereignty of the Cook Islands thereby confers a pragmatic advantage on its OFC and assists OFC users to negotiate much more forcefully with their antagonists. This is especially the case when contested property which is held through the OFC is outside the metropolitan court's country – so that it is virtually impossible to make direct seizures (*Australian Financial Review*, 25 November 1994; *The Economist*, 5 October 1991; *Forbes*, 27 April 1992).

The ability to create novel legal entities (such as asset protection trusts) for OFCs has been circumscribed in Norfolk Island, Guam, the Northern Marianas and (until recently) the Marshall Islands. They have comparatively little sovereignty and they have been manipulated by the United States (in the last three cases) and Australia (in relation to Norfolk Island) for their own ends. These two core states have largely defined their legal entities and their international relationships and powers. In the early 1980s the attempts by Guam to become a centre for 'treaty-shopping' holding companies were firmly rebuffed by Washington. The United States also acted against poorly capitalized and fraudulent offshore banks in the Northern Marianas, and then again in the Marshall Islands in 1984 after many of them migrated from the Northern Marianas to this somewhat more sovereign jurisdiction. On the other hand, in the mid- to late 1980s, the United States encouraged the creation of a flag of convenience for the Marshall Islands and also foreign sales corporations and relatively regulated and transparent offshore banking units in the least sovereign territory of Guam. The US government ambivalently attacks the low income tax rate on the Northern Marianas, but this continues to allow opportunities for companies as well as rich immigrants from the mainland. Australia has inhibited OFC developments on Norfolk Island – so that while there are still over 400 holding companies registered there, the number has declined from the height of 1,571 in 1971 (Treadgold, 1988, p. 219).

Low-tax regimes also have existed in the US-administered realms of Micronesia (from Palau to the Northern Marianas, the Federated States of Micronesia, and the Marshall Islands) but the profound ambivalence and uncertainty regarding the factual, legal and policy strength of two potentially competing powers – the island states and the US government (cf. Wentworth, 1993) – have discouraged tax haven development there. Even in a newly independent country such as the Marshall Islands (which was admitted to the United Nations on 17 September 1991), the recently developed OFC there is enfeebled by an 'Americanized' Mutual Legal Assistance Treaty with the United States which calls for exchanges of information about such matters as tax evasion (Langer, 1992), leaving it at a competitive disadvantage in relation to most other tax havens.

We have already seen how the US federal government repressed the aggressive entrepreneurship of unscrupulous people in the offshore sector. Past US governmental involvement has largely squelched the development of distinctive legal principles which would have served as the basis for many kinds of offshore activities, with the exception of OFCs in Guam (which serve US trade interests) and flag of convenience shipping in the Marshall Islands, which contribute to US military power (van Fossen, 1992).

Some have suggested, perhaps too cynically, that the region is so important to US defence strategy that Washington wants to maintain its dependency and opposes development strategies (e.g. of OFCs) which would provide revenues to lessen this dependence. In this light it is interesting to note that New Zealand raised no effective objection to the creation of OFCs in the Cook Islands (in 1981) and Niue (in 1993), nor did it enforce a burdensome Mutual Legal Assistance Treaty on these freely associated states – as the United States negotiated with the 'independent' Marshall Islands. Even the hostile 1994–96 New Zealand commission of inquiry into the use of the Cook Islands OFC by that country's large companies has been continually frustrated by Cook Islands financial secrecy laws, but Wellington has not tried to reduce the sovereignty which gives substantial protection to the clients of the Cook Islands OFC from such investgations. The United Kingdom has encouraged the orderly creation and emergence of tax havens in its colonies (such as the New Hebrides–Vanuatu), even colonies which had or have little or no strategic significance to it. The Bank of England has consistently sent its employees on secondment to many Commonwealth countries in the Pacific which are developing OFCs (Chaikin, 1992, p. 263). Whereas relatively reputable banks and institutions have been attracted to these jurisdictions, 'quick-buck artists' have been far more conspicuous in 'American' Micronesia's OFCs and have thwarted their effective conceptualization and planning.

Yet British OFCs lacking independence (the colonies, dependencies or territories of Bermuda, the Cayman Islands, the Channel Islands, the Isle of Man and Hong Kong) are often the most successful havens – in part because they do not appear to have as much freedom from outside standards as do more sovereign jurisdictions. Indeed, some years ago the British government commissioned a report and acted to regulate the conditions and practices of its Caribbean OFCs (Gallagher, 1990). In contrast to a superficially similar Australian report in relation to Norfolk Island (Australia, 1976), the British attempted to use this outside direction to improve the image and further the development of the OFCs concerned.

The connection between autonomy and successful OFC development is therefore ambiguous. Such extremely important OFCs as the Cayman Islands and Bermuda and such emerging centres as the British Virgin Islands, Gibraltar and the Turks and Caicos Islands continue eagerly to trade their possible or potential independence for fees and the advantages which come from being identified with a powerful, trusted and sympathetic core power. It is partly this continuing British help in OFC tactics and strategy which blocks the UN's initiatives towards self-government for these areas. Not least of the advantages which the United Kingdom has conferred on its OFCs is the image of stability, which is sometimes unavailable to independent states and which has assisted these British OFCs in attracting the business of many of the world's largest companies. Without full sovereignty they have clearly been able to create and organize internationally recognized companies and trusts.

More than politically independent states, the British jurisdictions have tended to enter into formal mutual legal assistance treaties which make minimal commitments to forbid the use of their OFCs for some serious criminal purposes (especially money-laundering in relation to narcotics-trafficking). Bermuda has gone furthest in agreeing, like the Marshall Islands, to exchange tax information with the United States. Most have strongly resisted assisting tax compliance and enforcement in any way.

The affiliation of so many prominent OFCs with the British state reveals its deeply divided character in relation to taxation and regulatory issues; the clash lies most clearly between the Inland Revenue and those political and bureaucratic interests allied with the City of London. The association with the City elite has given havens such as the Channel Islands and Bermuda, in particular, a genteel image which has placed them in the upper echelons of OFCs, with correspondingly high fees. The clients are paying in part for services rendered by extremely powerful, influential and respectable financial, corporate, legal, personal and governmental networks. The recent acquisition of the largest trust company in the Cook Islands in October 1994 by the Bank of Bermuda should be seen in terms of this hierarchy.

In these OFCs, promoters are likely to claim much greater concern for continuity of business with clients whose affairs they know far more thoroughly than their counterparts in cheap and notorious havens such as Panama and Honduras. Since they are powerful, cosmopolitan and reputable, they have tended to pre-empt threats to their OFCs from moral crusaders (e.g. those metropolitan politicians campaigning against money-laundering). They have been quicker than more provincial, less powerful and less respectable OFCs to accept some elements of their opponents' proposals – often through a kind of cooperative self-policing designed to stave off something worse, thereby preventing intense confrontation with metropolitan states. Being in a superior market position to begin with, they are better able to reject less respectable business. Yet large-corporation business is increasingly fragmented between a number of different OFCs, and many OFC entities are being used for one-shot transactions (cf. New Zealand, 1994, for Cook Islands examples of these). Even the most prestigious OFC insiders find it increasingly difficult to gain an adequate view of their clients' activities.

It seems likely, especially with the proliferation of OFCs, that there will be increasing pressure toward specialization, with each acquiring more character and a more distinct market niche. A clear hierarchy of PITHs has developed in the past decade, defined in terms of their clients, with the Cook Islands emerging as pre-eminent because it is the most closely associated with 'creative' structures for very large companies. Significantly, the Cook Islands is also the least independent of the major South Pacific PITHs. It has a relationship with New Zealand somewhat similar to that which the Channel Islands have with the United Kingdom, whereas Vanuatu, Western Samoa and Nauru are fully independent politically. However, it can terminate its relationship of free association with New Zealand at any time – so its effective sovereignty is rather high. The new PITH of Niue is similar to the Cook Islands in this regard. On the other hand, we must consider Norfolk Island and the other PITHs which have been severely limited or even destroyed as a result of their lack of political autonomy and the hostility of their metropolitan ruler. Independent sovereignty is not necessarily power, although it may be.

Sovereign territories have far greater ability to create and transmute corporate legal personalities and other legal devices (e.g. trusts) which are necessary for the OFCs' continual quest to create the most beneficial environment for capital. These nebulous and flexible entities help to disguise and to mystify the human drama relating to the assets which they hold. The attempts by metropolitan states to tax and regulate these forms have not been very successful – and these states themselves have started to enact laws to create similar entities which are able to escape taxation and regulation. Pacific Islands OFCs therefore contribute to the fragmentation of the metropolitan state. They create stateless financial areas outside the regulatory reach of metropolitan states, which these states may even attempt to formalize legally within themselves – perhaps in an attempt to reregulate them ultimately. This process reflects the supremacy of transnational capital. As stateless capital is increasingly generated in and attracted to the Pacific, Pacific Islands OFCs will abet its avoidance strategies and even invent new ones. The limitations, inefficiencies, contradictions and reactive qualities of the anti-tax haven measures of metropolitan states appear to promise sovereign and inventive Pacific Islands OFCs a bright future.

Notes

1. For a complete survey and analysis of Pacific Islands flags of convenience, see van Fossen (1992). Recently, with the termination of the US-administered Trust Territory of the Pacific Islands and the admission of the Marshall Islands to the United Nations, this archipelago has clearly gained far greater sovereignty. The decision of USLICO, the company which was then administering the Liberian OFC, to promote a new OFC in the Marshalls represented a parallel transition – so that by 1992 the Marshalls became (with the Cook Islands) the least sovereign jurisdictions in the first group, rather than being most sovereign in the second group.
2. The fact that so many of the largest companies which used the Cook Islands extensively in the 1980s, e.g. Bond, Ariadne, Equiticorp, Linter, Euro-national, Renouf, Judge, Industrial Equity, Bell Group, Bell Resources, are currently bankrupt or in greatly diminished circumstances illustrates that debtors may use OFCs' secrecy and tax provisions to obscure their financial positions and even claim substantial tax-free profits – thereby deceiving shareholders and creditors.
3. Allan Hawkins and six other former executives of the failed Equiticorp have been accused of a number of offences, including conspiracy to defraud the company of NZ$28 million. They allegedly attempted to launder the money through a 'loop' system which used the OFCs of Vanuatu (through First Pacific Finance, formerly Yeoman Ltd) and the Cook Islands (via Barely Grange, a shell company registered there). Court action commenced against them in September 1991. Hawkins was found guilty and is currently serving a prison term (*New Zealand Herald*, 1991–95, *passim*).

References

Atiyah, P.S. and Summers, R.S. (1987) *Form and Substance in Anglo-American Law: a Comparative Study of Legal Reasoning, Legal Theory, and Legal Institutions.* Oxford: Clarendon.

Australia (1976) *Report. Royal Commission into Matters Relating to Norfolk Island (Sir John Nimmo, Commissioner).* Canberra: Australian Government Publishing Service.

Chaikin, D.A. (1992) 'Investigating Criminal Corporate Money Trails.' In B. Fisse, D. Fraser and G. Coss (eds) *The Money Trail.* North Ryde, New South Wales: Law Book Company, pp. 257–293.

Crawford, J. (1989) 'Islands as Sovereign Nations.' *International and Comparative Law Quarterly*, vol. 38, pp. 277-298.

Gallagher, R. (1990) *Report of Mr. Rodney Gallagher of Coopers and Lybrand on the Survey of Offshore Financial Sectors in the Caribbean Dependent Territories.* London: HMSO.

Gild, M.E. (1990) 'Tax Treaty Shopping: Changes in the US Approach to Limitation on Benefit Provisions in Developing Country Treaties.' *Virginia Journal of International Law*, vol. 30, pp. 553–596.

Gordon, R.A. (1981) *Tax Havens and Their Uses by US Taxpayers.* Washington, DC: Internal Revenue Service.

Gray, J.M. (1985) 'Foreign Sales Corporations Replace Domestic International Sales Corporations.' *Harvard International Law Journal*, vol. 26, pp. 293–303.

International Monetary Fund (1995) *World Economic Outlook.* October. Washington, DC: IMF.

Langer, M. (1992) 'Marshall Islands.' In M. Langer (ed.) *Tax Havens of the World.* New York: Matthew Bender.

Naylor, R.T. (1987) *Hot Money and the Politics of Debt.* London: Unwin Hyman.

New Zealand (1994) *Papers Presented, by Leave, to the House of Representatives by the Member for Tauranga, the Honourable Winston Peters, on 16 March 1994*, vols 1, 2, 3. Wellington: House of Representatives.

Peters, B.G. (1991) *The Politics of Taxation: a Comparative Perspective.* Oxford: Blackwell.

Sampson, A. (1988) *The Money Lenders: Bankers in a Dangerous World.* London: Hodder & Stoughton.

Treadgold, M.L. (1988) *Bounteous Bestowal: the Economic History of Norfolk Island.* Canberra: Australian National University.

van Fossen, A. (1992) *The International Political Economy of Pacific Islands Flags of Convenience.* Nathan, Queensland: Centre for the Study of Australia Asia Relations of Griffith University.

Wentworth, L. (1993) 'The International Status and Personality of Micronesian Political Entities.' *ILSA Journal of International Law*, vol. 16, pp. 1–37.

Wilkes, O. (1987a) 'Sir Papadoc and the CIA.' Unpublished discussion paper, 22 April.

Wilkes, O. (1987b) 'Sir Papadoc and the CIA: Dirty Work in the Cook Islands.' *New Zealand Monthly Review*, vol. 301, pp. 3–7. Christchurch, New Zealand: New Zealand Monthly Review Society.

Winslow, D.A. (1990) 'Tax Avoidance and the Definition of Insurance: the Continuing Examination of Captive Insurance Companies.' *National Insurance Law Review*, vol. 4, pp. 743–824.

9

Majoring in Finance: Implications and Key Issues

Stephen Carse

Constitutional Position of the Isle of Man

The Isle of Man, which forms the subject of this chapter, is not part of the United Kingdom, nor has it ever been (see Figure 9.1). Rather it is a British Crown dependency with the Queen of England as its sovereign head. Basically, it has autonomy over its internal affairs, but the UK government assumes responsibility for its defence and external relations. The UK government also has ultimate responsibility for the good government of the Island.

The island's formal relationship with the European Union is governed by Protocol No. 3 to the United Kingdom's Act of Accession 1972, which took the United Kingdom into the Union. The Protocol affords Isle of Man visible goods the same treatment as accorded to trade between full members. Measures which relate to matters affecting the free movement of goods therefore apply to the Isle of Man. For all other purposes the Isle of Man is treated as a non-member. It pays no money into the European Union; it gets nothing out. The Isle of Man's economic access to the European Economic Area is the same as that to the European Union.

The General Economy

In its history the Isle of Man has been at various times a farming and fishing community, a mining area and a mass tourism centre, before taking on its present form as an offshore finance centre, a form which arguably can be traced back to the early 1960s when certain executive powers were transferred from the Lieutenant-Governor (the Queen's representative on the island) to the then Finance Board (now the Government Treasury). One of the

Figure 9.1   Location of the Isle of Man in the British Isles

first actions taken was to abolish surtax on personal incomes and reduce the standard rate of income tax.

The finance industry was given a particular boost in 1972 with the rescheduling of the Sterling Area (wherein the holding of sterling outside the United Kingdom was restricted, producing an inflow of capital from descheduled areas) and again in 1979 with the abolition of exchange controls. In 1973, when the United Kingdom became a full member of the European Community, the Isle of Man was given the special relationship in Protocol No. 3 to the United Kingdom's Treaty of Accession.

The 1970s was a relatively prosperous decade, with tourism performing well, with the growth of a finance sector, and success in the attraction of industrialists through the development of a strong package of incentives, so that by 1971 2,500 people were employed in the manufacturing sector.

The early 1980s was a traumatic time for the economy for several reasons, not least of which was a deep recession in the United Kingdom. The knock-on effect of the recession coincided with difficulties affecting the island's banking structure. The economy recovered strongly from the mid-1980s onwards as the UK emerged from recession towards economic boom and as personal and corporate awareness in respect of financial methods and instruments grew. By the time of the next downturn in the economic cycle, the island's economy had grown sufficiently more robust and sufficiently diversified to avoid technical recession.

After taking into account inflation, Isle of Man GNP increased by 61 per cent between 1985/86 and 1993/94 (the most recent year for which statistics are available) to £584 million (£8,210 per capita). Real national income has more than doubled since 1970. The major income-generating sector is finance, which (when defined as insurance, banking, other financial institutions, property owning and management, and related business services) contributed £198 million (35 per cent of total income) from Manx sources. Other sectors include professional services (15 per cent), which includes advocacy and accountancy, manufacturing (11 per cent), tourism (6 per cent) and agriculture (2 per cent). Transport and communications provides another 8 per cent, construction 6 per cent and retailing and wholesaling distribution 9 per cent.

Clearly there have been several major changes to the island's economic structure in the past twenty years. Figure 9.3 and 9.4 later in the chapter illustrate the success of the finance industry. They also show a marked decline in the relative importance of tourism, and indeed of most other sectors. However, it is not that these sectors have not expanded, rather that their growth has been outstripped by that of the financial sector. The current predominance of the finance industry actually dates back much further than the mid-1980s. It was as long ago as 1973 that the industry took over from manufacturing as the leading income-generating sector.

Though of course we talk in terms of sectors, no one sector stands independent of any other. Moreover, each sector comprises numerous, often unrelated, activities. The financial services sector, for instance, now includes 61 licensed banks, 121 collective investment funds and 149 captive insurance companies. It also includes 14 life assurance companies, plus pension fund managers, trust and company formation firms, ship management and so on. The manufacturing sector comprises some 250 companies, predominantly operating in the light engineering sector but with important operations also in food processing and textiles, for instance.

Given the close relationship between economic and population growth, it was not surprising that the buoyant economic climate in the latter half of the last decade produced an increase in the island's population level. The freedom of movement of individuals between the island and the United Kingdom means that net migration will occur in the direction of the more successful economy. Isle of Man resident population increased from 64,282 in 1986 to 69,788 in 1991 and is currently estimated to be around 72,000. Immigration occurred predominantly in the economically active age range, improving the island's demographic profile by reducing the proportion above retirement age. The age structure of residents in the 1991 census was recorded as 19 per cent children, 59 per cent of working age and 22 per cent retired. And whereas the predominant age grouping twenty years ago was the age band 60–69 (accounting for 14 per cent of the total population), now the number of retired people, in both absolute and relative terms, is in decline and the number of economically active individuals increasing. The largest age group now is of people in their twenties (14 per cent) while the proportion of people in their sixties has fallen to under 10 per cent.

The inflow of new residents in recent years effected significant changes in the island's employment structure. The 1991 census revealed that 31,829 people were in employment in the Isle of Man, some 6,000 more than 10 years earlier. The economically active population numbered 33,189. Immigration and demographic changes have boosted this figure to over 34,000. The census results revealed a major increase in employment in the finance industry, from 1,515 in 1981 to 4,353 in 1991. When combined with the professional services sector, the total reached 9,791, or 31 per cent of the workforce in employment. Current estimates of the finance sector alone put employment there now at 5,800. The other major sectors in terms of employment generation are manufacturing (11 per cent), construction (11 per cent) and tourism (10 per cent), further reflecting the broad base of the Manx economy.

Government receipts in 1994/95 totalled £200 million, of which £98 million (49 per cent) was raised through indirect taxation and £95 million (47 per cent) was generated by income and other direct taxes. On the other side of the account, net expenditure for the same year totalled £198 million. In the financial year 1995/6 departments with the largest budgets are Health and Social Security (£86 million) and Education (£42 million).

In all respects, therefore, the Isle of Man is a much bigger economic entity now than ever before. It is also much fuller, more complete, with economic activity in more and more areas. It will never be a 'microcosm' of the United Kingdom. It will never, for instance, have a much bigger manufacturing sector, if only because the size of the Island's resident population will never be large enough to persuade manufacturers to service local consumer demand from manufacturing units on the island. But the island is slowly filling in in other areas: in shipping, in higher education provision, in computer services and in numerous other fields.

Reflecting the general economic recovery, unemployment has fallen in 1995, to stand currently at 4.2 per cent. The success of the economy in withstanding the ravages of recession elsewhere, particularly in the United Kingdom, is shown by the fact that the unemployment rate never rose above 5.5 per cent in the early 1990s. Indeed, unemployment has been at or below this rate continuously since May 1987, and fell to only 1.3 per cent in May 1990. Undoubtedly the continuing geographical diversification of the island's customer base has been a key factor in avoiding recession. Nevertheless, in view of the plethora of economic links, economic conditions in the United Kingdom will continue to affect the Isle of Man economy as illustrated in Figure 9.2.

Global Factors in General Economic Development

One of the most important developments in the past decade has undoubtedly been the growth in the power of the financial markets and its major players. The liberalization of the markets and of world capital movement not only has

Figure 9.2 (a) Annual rate of inflation (%)

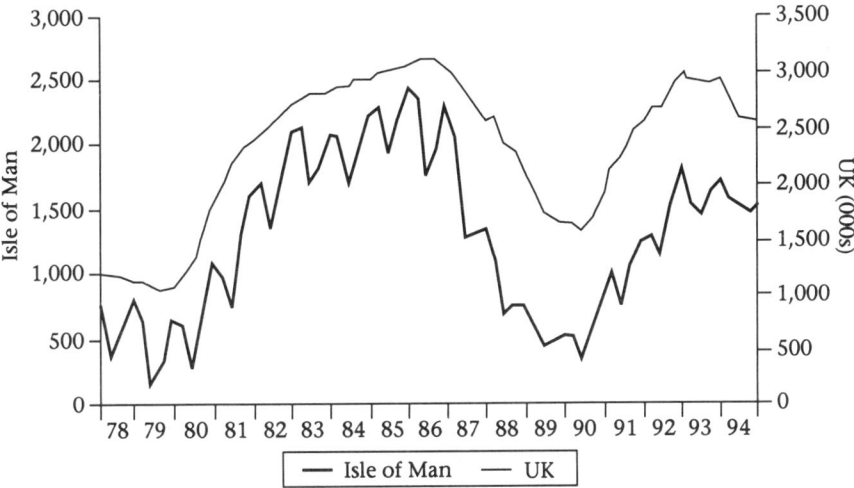

Figure 9.2 (b) IOM and UK unemployment

rendered many traditional economic policy instruments redundant, but also has served to make the financial markets the ultimate judges and imposers of discipline on government economic policy. The implications of all this have been illustrated clearly in recent years but have perhaps yet to be fully appreciated by governments still trying to grasp the extent of the resultant loss in their economic sovereignty.

Currently some half of world output and income is generated by the economies of the United States, Japan and Germany. It is inconceivable that the health of Western economies will continue to be other than primarily affected by the condition of these three countries. Yet their dominance in shaping the global economy is gradually diminishing with the expansion of the newly developing nations, particularly those in Asia. These economies have grown collectively at annual rates of around 6 and 7 per cent in recent years despite recession elsewhere.

The competitiveness of the developing economies stems naturally from low absolute and per unit labour costs, and is aided by the potential market size of both the domestic and neighbouring economies. With the benefits of being able to produce at low cost being realized by agreements under the auspices of GATT (and, in future, the World Trade Organization), it is not surprising that a drift of manufacturing investment to the East is evident. While market location and certainty of supply will always be critical factors affecting locational decision-making, it is the increasingly prime requirement to minimize labour costs that will ensure the continued growth of the Asian economies.

It is nothing new to Western industries to be under pressure from emerging nations. The scale of the concerns over the present and future competition reflects the coinciding of that heightened threat with recession and the seemingly ever-increasing rate of labour-saving technological innovation and diffusion. The need to minimize costs and maximize productivity among companies involved in export markets is now considerably more marked, and the liberalization of world trading generally has left few sectors unaffected by the prospect of foreign competition. The portents for Western unemployment are clear. However, an economy of successful producers is also an economy of increasingly wealthy consumers, and therein lies the key to the success of the leading economies. While it is difficult to see Western exporters competing successfully in selling standardized consumer products, there will be strong and growing demand for the services of developed economies in the provision of the physical infrastructure, in engineering, in telecommunications and in financial services. At a later stage in their development the newly industrializing nations will have greater demand for less standardized, more differentiated, consumer products. The more successful economies will be those whose companies are geared towards the competitive supply of such goods and services, not just to the emerging countries but also to each other.

Faced with increasing competitive pressures there is already evidence of the reaction from Western governments:

- the introduction of measures to 'free up' inflexible labour markets and reduce labour (particularly non-wage labour) costs in order to improve competitiveness in those markets and attract inward direct investment;
- competing for inward investment via reductions in direct taxation and enhanced financial incentives;

- formation of regional trading blocs to further regional trading but also to enhance trade protection and encourage trade between member nations.

To these may also be added another reaction, though one borne out of market forces rather than government action, namely the further specialization into services and away from manufacturing.

Clearly, the competition facing Isle of Man companies and facing the Isle of Man government in its efforts to attract direct investment will continue to intensify. Yet the global trends alluded to here are best seen as background factors affecting policy decisions. For it remains the case that, whatever the level of global investment and trade, the Isle of Man needs only to secure a minute fraction of that business to continue on its development path.

The Financial Services Industry

Importance

The financial services industry in the Isle of Man is responsible for:

- over a third of national income (40 per cent if related professional services are added);
- almost three-quarters of the island's company income;
- 1 in 6 of all jobs on the island; and
- around 80 per cent of the total company tax take.

On any account, therefore, it is an absolutely crucial part of the Manx economy.

The finance sector is often referred to as 'new'. Though this is certainly true in comparison with agriculture and with tourism, it is less true when the

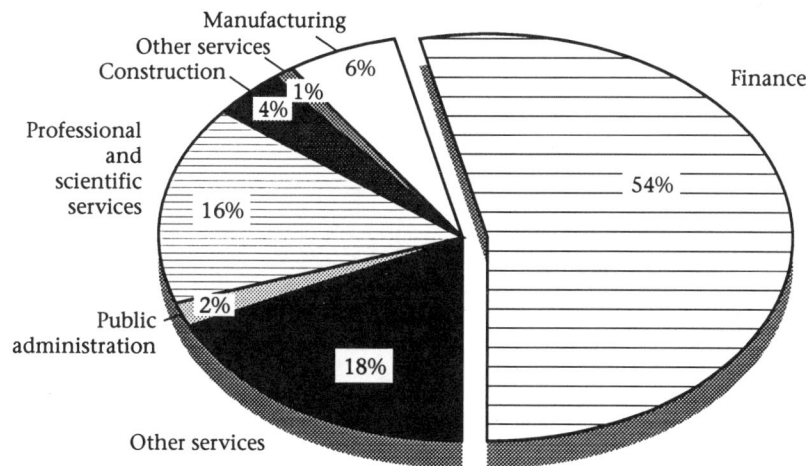

Figure 9.3 Sectoral contributions to growth 1983–1993

1972/73

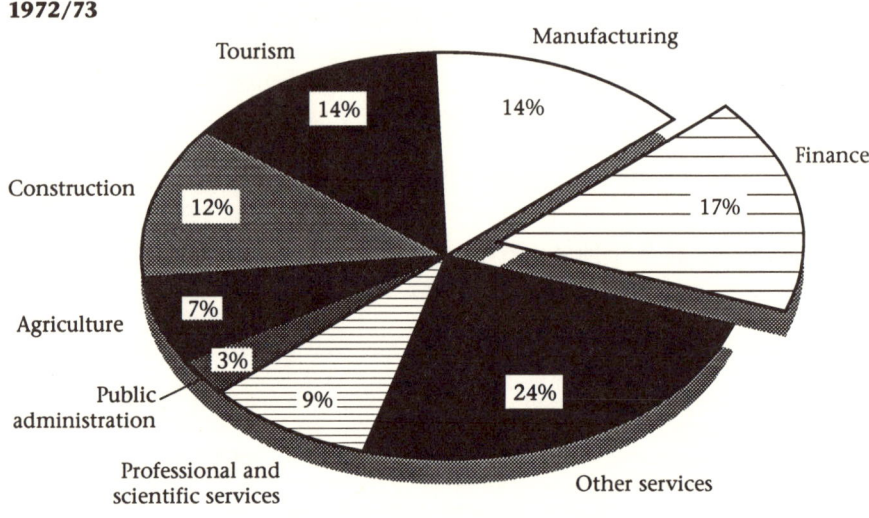

1992/93

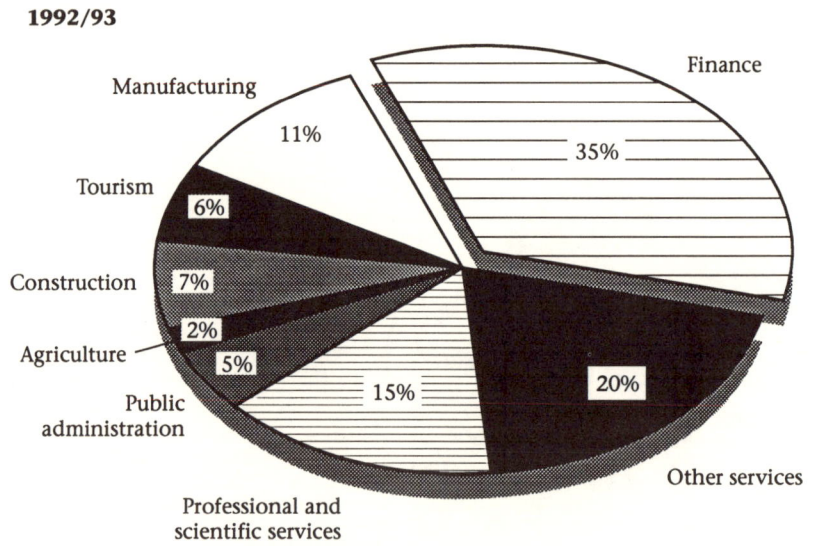

Figure 9.4   Sectoral composition of national income 1972/73 and 1992/93

comparison is made with manufacturing, a sector which itself began to take on true significance only in the 1960s. The history of the finance sector itself now goes back over thirty years. As noted already, it has been the major income-earning sector of the economy since as long ago as 1973.

In the past ten years alone, the finance sector has accounted for over half of the economy's expansion. Figure 9.3 shows the growth rates of the financial services sector against the rest of the economy since 1983. Figure 9.4 reveals how over the same period the structure of the economy has altered, producing criticisms from certain quarters of having 'too many eggs in one basket'.

The 1980s revolution

The 1980s was a decade of momentous change affecting the world of international finance. The principal factors effecting this change were:

- the removal of exchange controls, initially in the UK but then followed by other nations;
- the liberalization, deregulation and internationalization of financial markets;
- technological advance, especially in communications;
- innovation in the provision and form of financial products, e.g. derivatives, spectrum filling, secondary instruments (securitization), risk-covering instruments, etc.;
- new markets such as the London International Financial Futures and Options Exchange (LIFFE); and
- development of niche markets.

What were some of the effects of this change and innovation?

First and foremost, it has greatly increased the amount of financial services business undertaken worldwide. It has widened consumer choice in the conduct of financial affairs and it has pitted financial firms against each other in their quest to attract new customers. It has determined that many of the features which traditionally differentiated banks from other financial institutions are disappearing. The major banks now perform a much wider range of non-banking transactions, and similarly, non-banks are offering services which were traditionally the prerogative of the banks. The banks' traditional retailing activity of cash provision and deposit services is also being attacked by automation, by the development of machines which do not require a branch, something which will increasingly make large branch networks redundant.

In the same way that the differences between the banks and the other financial institutions such as insurance companies and investment businesses are becoming blurred, much the same is beginning to happen with the distinction between finance and non-finance companies. In the United Kingdom, for example, a number of retailers have experimented with financial innovation on a large scale and have in certain respects, such as in the provision of credit cards, become financial institutions. Marks & Spencer has a banking licence and markets investment products. More recently, and more dangerously, large, typically multinational, companies have seen their finance departments evolve from operating purely defensively in a way designed to minimize and cover risks, to become more aggressive players of the international financial markets through the use of options, futures and other derivatives. Although ostensibly still financial departments of conglomerates, these units are operating increasingly as investment arms.

For national governments, changes in the financial world have had very important effects on the extent to which they are now able to control their macro-economies. First, the changes have added hugely to world liquidity over the past decade. Second, some of the traditionally important macro-economic instruments have been made redundant. For example, national credit controls are no longer effective, given that borrowers can obtain funding from foreign markets. Domestic interest rates can no longer be manipulated in response to domestic demand requirements in view of repercussions they now have on the global capital markets as a consequence of market liberalization and internationalization. Indeed, it can now validly be said that the financial markets have become the ultimate arbiters of government performance and the ultimate discipliners of government policy.

For employment, the sum of all these effects is to increase the labour requirement in the financial services industry worldwide but at a rate less than commensurate with that of the increase in the volume of business itself. The same conclusion might be brought to the level of individual offshore jurisdictions.

Outlook

Within this general framework of the development of international financial services worldwide, there is an optimistic outlook for the Isle of Man. It is generally believed that world conditions will be favourable for this sector with demand for such services continuing to expand as a consequence of factors including:

- further liberalization of economies and financial markets;
- economic development of emerging economies;
- increasing disposable incomes and wealth;
- increasing corporate and individual awareness of the use of financial products;
- removal of barriers to the free movement of goods, services, capital and labour; and
- innovations in financial markets and products.

The business is certainly there to be won. A 1993 survey conducted by Chase Manhattan Private Bank estimated that over 20 per cent of private wealth worldwide ($2.1 trillion) is invested 'offshore'; that is, with an institution other than the individual's domicile. A similar proportion of wealthy individuals hold their assets offshore. Interestingly, it was further calculated that while one-third of offshore holdings come *from* Europe, almost two-thirds are held *in* Europe.

The Isle of Man government over the past few years has developed its own central planning assumptions, which provide a view on the short- to medium-term outlook for the economy. The assumptions as they currently stand project average real economic growth of around 5 per cent over the rest of the 1990s and a total employment level by then of something approaching 36,000, compared to under 32,000 at the time of the 1991 census.

It is assumed that the finance sector will grow at over 6 per cent a year and thus contribute over half of the total expansion in terms of national income and one-third of the job growth.

The sector's 'needs'

As noted earlier, the financial services sector on the Isle of Man contains a multiplicity of activities, including deposit-taking, fund management, life assurance, captive insurance, trusts, company formation, ship management, legal services and accountancy. Within these areas there are also a multiplicity of activities and specialisms. More so than within the other sectors on the island, there are 'linkages' between the different subsectors and companies within the sector; that is, the work of one frequently requires the involvement of another. Accordingly, the island now has a well-developed financial and professional infrastructure, which serves to provide general support for the industry. Perhaps the most important, and self-sustaining, strength of the sector derives from what might be labelled 'economies of localization'; that is, the development of the sector so that it now supplies such a width and depth of services and a pool of expertise as to increase the attractiveness of the Isle of Man as a location for additional entrants and for additional business.

Such a network of support and complementary activity is crucial for the enduring success of any offshore centre. Taking banking as an example, one could list the following services as complementary to both the local domestic economy and the offshore industry:

- cheque accounts
- investment/Treasury accounts – call and notice
- currency accounts
- foreign exchange – spot, forward, swap rates, etc.
- letters of credit
- negotiation/collection of remittances

- money transfer – cross-border

while there are a whole number of other services that will be needed to support the offshore sector including:

- electronic banking with full interface/networking facilities
- Treasury facilities
- correspondent bank links worldwide
- international direct debiting (life sector)
- cross-border small-value payments
- fund management
- custodial and administration services
- trust services
- managed bank services
- structured finance, i.e. support with major acquisitions outside the home country
- international trade finance structures, e.g. in the Isle of Man, VAT 'triangulation'
- legal services.

More generally, one could list a whole range of conditions that need to exist for any offshore centre to be successful. These would include:

- 'AAA' country debt rating (Standard & Poor);
- political and economic stability;
- low-tax regime including personal taxation;
- space and encouragement for growth;
- established excellence in banking, trust, insurance and other financial services;
- the existence of excellent professional support services and an established infrastructure;
- low set-up and operating costs;
- high quality of life;
- well-educated, available labour force;
- low-cost, high-quality office accommodation and housing availability;
- situation in a time zone central for European, Far East and US markets;
- OECD jurisdiction;
- modern communications;
- absence of exchange controls;
- strategic location;
- high standards of regulation and stringent investor protection;
- a responsive and supportive government.

The Isle of Man meets all these requirements, and more. If the intention is simply to service the needs of a 'parent' country, the developing offshore centres might be able to succeed by achieving on a limited number of the criteria. But if they aspire to becoming a truly international centre like the Isle of Man, with a worldwide customer base, this checklist will need to be addressed.

Towards a Strategic Approach to Determining Policy on Development

The process

For a host of reasons, ranging from a lack of internal resources to a resignation in the face of the overwhelming power of uncontrollable external forces, small-country governments have typically avoided what might be called a strategic approach to economic policy development, settling instead for a broad policy objective of improved living standards and a policy stance that is generally reactive to events, opportunities and threats.

But now they are becoming much more sophisticated in their outlook, much more cognizant of the benefits of a more strategic, more coherent and more planned outlook, wherein objectives are clearly stated, forward assessments made of strengths and weaknesses, opportunities and threats, and an economic policy then mapped out.

Objectives

The key statement of the Isle of Man government's economic goal is 'To pursue manageable and sustainable growth based on a diversified economy, with the aims of: (a) raising the standard of living of the whole population, (b) securing continuous future prosperity, and (c) providing the resources needed to sustain public services'. These are fine and uncontestable sentiments. It is the means of achieving the objective that provide the policy dilemmas.

Limitations

To begin with, there are the difficulties imposed on small countries simply as a consequence of their size: economic dependence on larger economies, a general absence of scale economies, etc. Then there are the limitations which can arise from existing economic arrangements. In the case of the Isle of Man these might be seen to include:

- its monetary union with the United Kingdom;
- its customs union with the United Kingdom;
- its statutory balanced-budget requirement;
- its relationship with the European Union.

The principal limitations on the kinds of policies that the Isle of Man government can adopt are a consequence of the monetary union with the United Kingdom. The Isle of Man has no independent currency and no independently set interest rates. Accordingly, it does not have recourse to policies aimed at affecting either its currency exchange rate or its base lending rates. Furthermore, fiscal policy is constrained since as a consequence of the voluntary customs union with the United Kingdom (via its

Customs and Excise Agreement) there is limited freedom of action in respect of indirect taxation.

There is a statutory requirement that government has to plan for a budget surplus, or at least to avoid a budget deficit. There is no option of using a budget deficit as a countervailing measure against any economic downturn. But that is not to say, of course, that one cannot use the absolute level of government spending, or indeed the tax system, as a means of affecting the economic climate. But it does prevent extremes. As a consequence, the Isle of Man, as a rarity among nations, has no national debt. Indeed, it has a general reserve fund currently valued at over £90 million. Neither, of course, does it have any equivalent to a public-sector borrowing requirement (even if from time to time there is a need to use income from the reserve fund to finance expenditure).

The final principal constraint listed above on the island's policy actions is that imposed via obligations arising under arrangements with the European Union via Protocol 3 of the United Kingdom's Treaty of Accession. The European Union imposes on the Isle of Man in several ways:

1. Most obviously, some Union decisions apply directly via the Protocol, for example on customs and quantitative restrictions.
2. Other Union decisions do not apply to the island but are adopted by virtue of an agreement with the United Kingdom or because UK legislation extends to the island. The general rates of value-added tax are a good example here.
3. Some Union decisions apply standards or conditions on trade within and into the Union and these apply to the island only to the extent that island companies wish to trade within the Union.
4. In certain instances the island adopts EU initiatives not because it is obliged to but because it sees EU standards, for example in respect of bathing water, as standards (perhaps the minimum standards) that should be met.
5. Union decisions affect the island via their general impact on the world order; for example, in respect of the liberalization of trade.

Clearly, then, many of the more significant impacts of the Union on the island do not arise from the island's formal relationship as denoted in Protocol 3.

When we examine all these restraints on the possible actions of the government, we may note that they all arise through choice. None of the restrictions is irreversible. They are maintained because it is believed that they constitute the best possible position for the Isle of Man. In other words, these positions might restrict courses of action but they are not seen as constraining economic development.

Furthermore, to conclude that the Isle of Man government has little power to influence the nation's economic fortunes is to ignore the potentially powerful range of instruments which are available:

- taxation
- government grants and loans
- general government spending – the overall level and type of expenditure

- marketing
- licensing and regulation
- commercial legislation
- work permits
- labour market policies
- cost controls – direct controls (e.g. government-controlled rent and rates), and indirect controls (e.g. public-sector wage levels).

Key Issues for Governments of Offshore Centres

Diversification versus specialization

Arguably one of the most important policy decisions that has to be made concerns the alternatives of development through diversification or specialization. Classical economic principles espouse the latter, particularly for open economies. But there is an attraction for the risk-averse government in seeking greater diversification, having economic activity spread across a number of different sectors, and across a range of activities within each sector. In accepting this latter path there is an implicit choice being expressed for more manageable, yet more sustained, growth in preference to more erratic, though at times more spectacular, expansion.

Most governments will feel much more comfortable with opting for a diversified economic base. The Isle of Man government is no exception. It maintains a number of policies which serve to encourage growth outside the finance industry. In particular, it offers grants and loans of up to 50 per cent and five-year tax holidays to manufacturing concerns. Like all governments, it subsidizes its farming industry (farmland accounts for four-fifths of the island's land mass). It supports tourism with substantial grants for individual projects and accommodation upgrading. And its maintenance of present customs arrangements has the primary objective of sustaining its manufacturing and agricultural sectors by ensuring a continued absence of physical customs barriers between the island and the United Kingdom.

Of course, it is in any event too simplistic to talk of sectors as if they are homogeneous entities. The reality is of great diversity within each. Accordingly, even where finance is the predominant sector in a country, it is crude to conclude that the fortunes of each component rise and fall in tandem. To take just one example from recent experience, falling interest rates in the British Isles may have been bad news for deposit-takers but quite the reverse for investment managers.

Regulation

There will always be debate over regulation: whether it is too light or too heavy-handed; whether it assists the Isle of Man to win business or helps lose it; whether it should be held at current frontiers or extended to presently unregulated areas. It is yet another of those fine balancing acts.

Elsewhere, generally the debate is fundamentally concerned with consumer protection. In offshore centres it is equally to do with protecting the reputation of the jurisdiction and its financial services industry, and is thus more related to issues of fraud, tax evasion and money laundering.

Regulators cannot reasonably be expected to guarantee 100 per cent success in respect of consumer protection. This could be achieved only by the imposition of a huge cost in terms of damage to the competitive efficiency of the financial services industry. And there is not much point in arrangements which provide absolute protection against risk in relation to financial transactions if this means that the cost of those transactions subsequently puts them out of reach.

Neither can regulators prevent all kinds of abuse of the financial system. But by seeking to ensure that institutions are run by 'fit and proper' persons, 'know their customers', and have adequate systems and procedures in place to minimize abuses, a strong regulatory system represents a very powerful deterrent.

The Isle of Man is known as having a well-regulated industry and a system that meets (and in places exceeds) international standards. It has a depositor protection scheme equivalent to UK rules, but covering foreign as well as sterling deposits. The Isle of Man government believes that there is a benefit to be had in maintaining its high supervisory and regulatory standards, and that this quality confers benefits on all who operate from the country. It does not want the laundered funds, the harbouring of incomes from illicit dealings. It *does* want to conduct real investment business for genuine clients. The demand for such services is great and growing worldwide. The need for regulation in the area of company formation and trust formation is, I think, well recognized everywhere, and future extension of supervision in these areas will need to strike the right balance too.

There will probably never be total agreement over the level of regulation and supervision, but perhaps there would be more agreement on the principle that requirements should be broadly equivalent across institutions undertaking similar activities. With the blurring of the distinction between the business undertaken and the products offered by the different financial institutions it would seem logical that single regulatory bodies ought for the future to be designed to cover all areas of the industry. In this regard the structure and remit of Malta's own Financial Services Centre seems eminently sensible.

Competition

Competition between low-tax areas, especially as offshore finance areas, is becoming more and more intense, while new areas seem to be coming on stream almost monthly. Though authorities in the Isle of Man pay particularly close attention to what is coming out of the Channel Islands, it is recognized that all are potential competitors and the actions of all are closely monitored. However, if they just keep an eye on the small jurisdictions they risk losing sight of the fact that many large countries are now setting themselves up, fiscally and legislatively, as low-tax areas for personal and corporate operations – none more so, perhaps, than the Isle of Man's neighbour the United Kingdom. Indeed, the Chase Manhattan survey noted earlier estimates that half the personal wealth held 'offshore' is in fact lodged in Switzerland and the United Kingdom. Certainly the United Kingdom Inland Revenue now appears geared up to devising fiscal changes designed to reduce the attractiveness of the British offshores, as exemplified by Budget changes to the United Kingdom's controlled foreign company legislation.

We are likely to see much more of this kind of action being taken unilaterally by national governments. Alongside the heightened competition, we have all become aware of the attention now being paid by individual countries and regional blocs to the matter of perfectly legal capital movements into or via the small offshore finance centres. This could be an increasing source of tension for the future. Doubtless there will always be moves emanating from bodies such as the European Union and the OECD aimed at assessing the true value of offshore jurisdictions and assessing ways in which the flow of capital and business into such centres might be reversed. In order to offset any misguided actions it is important that offshore centres nurture good relations with the international bodies, and impress upon them the economic benefits of their financial services activities.

Taxation

There can be no denying that it is the Isle of Man's low-tax environment which has been responsible for its economic development. While one cannot deny the importance of excellent communications, political stability, an innovative and efficient industry infrastructure, quality of service, low operating costs, and so on, it would be disingenuous to suggest anything other than that low taxes are a prerequisite for any offshore jurisdiction.

The Isle of Man has always considered itself innovative in its direct-tax policy. It now has an extensive range of tax possibilities for companies:

Non-resident

Either:

1. Manx incorporated company declaring itself to be a non-resident, or
2. foreign incorporated company not managed or controlled in the island, but operating through a permanent establishment.

Liability to non-resident income tax at 20 per cent would arise on Manx-source income (subject to any concessionary exemptions).

Resident

Either:

1.   Manx incorporated company, or
2.   foreign incorporated company which is managed and controlled from within the island.

Depending on ownership and nature of business:

- insurance company paying no income tax;
- insurance company paying income tax;
- managed bank paying no income tax;
- international loan concession for banks with a 2 per cent rate of tax;
- tax holiday break of up to five years;
- exempt company paying a fee;
- international company paying income tax of up to 35 per cent;
- fund management company liable at a rate equivalent to 5 per cent;
- company paying income tax at 20 per cent;
- company liable to 20 per cent tax but distributing the whole of its income without non-resident tax being deducted;
- petroleum production tax at 25 per cent.

The island's direct-tax policies for consideration for the future include:

- maintaining a direct taxation policy which remains attractive in relation to other jurisdictions;
- encouraging genuine operations to be set up in the island;
- encouraging headquarter operations to be set up on the island;
- seeking to extend arrangements to other countries where there is a clear business opportunity arising out of such an arrangement;
- seeking out niche markets and creating the right level of tax and regulatory control to encourage expansion, e.g. shipping and pensions.

One issue concerning all governments involved in attracting foreign investment is of course the danger of tax degradation: the competitive downward spiral in levels of direct tax and the competitive narrowing of the direct tax base. Firm public-spending control and a buoyant economy with strong tax receipts mean that the Isle of Man is in a strong position to compete on these terms. Given the healthy condition of the island's public finances, it seems strange that the associated low taxation can be deemed to be tantamount to 'unfair competition'. The Isle of Man government does not see future tax levels as the prime ground for competing for future market share. It is quality of service and regulation that provide the true arena.

Europe

On the subject of the European Union it is natural that the Manx authorities should examine closely everything and every rumour that comes out of the Commission. The Isle of Man government is quite relaxed over the true size of the anti-offshore threat likely to emanate from that quarter in the short to medium term. That is not to say that anything the Union does in the financial and monetary field would be of no importance. Certainly the island must pay great attention to ongoing moves for, for example, an EC-wide withholding tax or an EC-wide minimum corporation tax, although in these instances actions taken could present opportunity to the Isle of Man rather than threat.

The prospect of a single European currency is more realistic, and the implications for the Isle of Man will vary depending on whether the British government adopts the currency itself. The logic of a single trading currency, by reducing transaction costs and exchange rate uncertainty, is inarguable. But I fear for the economic consequences of a move which takes away so much power from national governments and restricts their ability to counter recession and unemployment in their own economies.

The adoption of a single currency will have the effect of reinforcing the tendency of members to trade with each other. To this extent it will be harmful for the island for sterling to stay outside the system. In many ways the 'single-currency gang' ought to be viewed like an 'urban street gang': not in itself desirable but potentially much more harmful not to be a member of. I think it would be much more beneficial for the Isle of Man if its currency is that of the European bloc rather than just that of the United Kingdom, even if the advent of a European currency does reduce the currency exchange profits for the banks, narrow currency portfolios, and involve everyone in translation costs.

Legislation

Aside from taxation policy, arguably the biggest influence a government can have on forging economic success is in the form and timeliness of its commercial legislation. It is a government's key duty to create an environment that is conducive to commercial success and accommodating to quality practice. Certainly the government has to react quickly to the ever-changing marketplace, with fiscal and legislative changes of its own in order to assist resident companies win business, and to attract in new companies.

On the whole, the legislative system on the island has performed creditably. In the past two or three years alone commercial legislation has seen the introduction of managed banks, of the reduced tax rate for fund managers, of bareboat chartering for shippers, and of the international company and international limited partnership. There is legislation in the pipeline to enhance business opportunities in the areas of trusts, insurance redomiciliation and limited liability. Recent years have seen the introduction of regulation of investment business, while legislation is timetabled in the next two years on regulation in respect of corporate administration, fiduciaries and timeshare operations. The aim in the programme is to maximize the potential for economic success without compromising regulatory standards.

Government marketing

Government clearly has an important role to play in marketing and promotion. The island's larger companies, particularly subsidiaries of international parents, are capable of looking after themselves. But there are a large number of companies of insufficient size to conduct promotion and marketing themselves. Moreover, even the larger companies will benefit from a generally prosperous and successful economy. And everyone benefits from promotion which highlights the awareness of the island among global investors and savers.

The success of the economy in recent years has created something of a dilemma in how to react to phases of expansion. In times of 'boom', when new companies are establishing a presence and existing companies are taking on more and more staff, government has been faced with two options:

- should it 'make hay while the sun shines' and in other words be thankful for all the growth it can get; or
- mindful of its commitment to 'manageable' growth, should it alter its policies and marketing in an effort to moderate the expansion, perhaps at the expense of lost opportunities and investment?

It was reaction to the speed of expansion in the period 1986–88 that ultimately led the government to curtail its own expansion policies then and cease marketing. But once having turned off the growth effort it was difficult to rekindle the outside interest.

Population and manpower

The Isle of Man stands in a minority of small offshores in having no residence controls at present. But the enaction of any residence control system must proceed alongside a reinforced message that the Isle of Man is 'open for business' and is working to economic objectives which openly admit to the need for immigrant labour. The Isle of Man as a location on the basis of the potential for expansion it offers to companies is a key attraction. Enabling legislation is scheduled but currently it is felt that the achievement of economic goals requires a larger population.

Conclusion

It is frequently the case that small countries have few natural physical endowments from which economic development can stem. They often suffer too from endemic disadvantages in respect of small size. For such countries, economic success has to be 'engineered'; governments have to use other weapons under their control to make their countries wield a comparative advantage in one or more areas of economic activity. One such weapon is taxation. Another is accompanying commercial legislation.

This is the path chosen by increasing numbers of small, often insular, countries. It was the path chosen by the Isle of Man in developing its financial services industry. It is a course which is fraught with policy difficulties and one which is permanently vulnerable to external forces and circumstances in a very competitive environment.

Though the portents for growth, even in an increasingly clustered market-place, are sound, offshore centres must ensure compatibility (at least) with larger nations in respect of licensing, regulation and supervision. Hostility towards offshores can also stem from ignorance of their role in facilitating capital movement and investment, and from notions of 'unfair competition' from nations unable to offer low tax rates. Such claims are illogical. Taxation will remain a genuine weapon in international competition for the attraction of business and investment in all economic spheres. But it will pay finance centres to be seen to be more than just low-tax areas and to have well-developed, diversified economies, with large numbers of companies with a real trading presence.

From a purely government perspective, the days when policy could be made reactively are in the past. Governments have to be fully aware of the international forces that will dictate future success or failure. More strategic assessments and policies are now required.

10

The Impact of International Companies on the Economies of Small Islands: a Case Study of Bermuda

Brian Archer

This chapter summarizes the results of a series of studies carried out in Bermuda mainly during the period 1985–94 to measure the impact of international company business activity on the economy of Bermuda. These studies were undertaken on behalf of Bermuda's Ministry of Finance, with the full support of the international business community.

The specific objectives were to measure the amount of income, employment and public-sector revenue generated in Bermuda by the activities of these companies and also their effect on the balance of payments. The work was carried out using input–output models constructed specifically for each study. Major surveys of business activity in Bermuda were carried out in conjunction with the Bermuda Statistical Department for each of the base years 1985, 1987 and 1992 – years selected by the Bermudian government. Intervening years were assessed on the basis of sample surveys plus the use of current data from other published and unpublished sources – relating to both the public and the private sectors. The 1985 base year model contained 19 sectors and included data from 259 establishments; the 1987 base year model contained 33 sectors and included data from 272 establishments; the 1992 base year model also contained 33 sectors and included data from 277 establishments. In all three base years, the data covered at least 90 per cent of total economic activity in the country.

Input–Output Analysis

Input–output analysis is basically a method of examining an economy and simulating the effect of changes. All sectors of the economy – industrial, commercial, etc. – are shown as sellers and buyers. The sales of each sector to other sectors are shown as rows in a matrix and their purchases from other sectors are shown as columns. At the same time, their sales to final demand (consumer expenditure, government and exports) are shown outside the matrix as a series of vectors, and their purchases of factors of production (wages and salaries, profits, etc.) are shown as rows beneath the matrix.

The effects of any changes (primary and secondary) are obtained by inverting the (I–A) matrix and then post-multiplying by a vector of these changes. The technique is explained more fully in Archer (1973).

Bermuda: the Background

Bermuda, a British colony, is a cluster of some 100 small islands, 20 of which are inhabited, situated in the Atlantic Ocean about 570 miles east of Cape Hatteras and about 6½ hours' flying time from London (see Figure 10.1). The six principal islands are joined by bridges and a causeway to make up a total land area of less than 21 square miles. The population currently exceeds 60,000. The Bermudian dollar exchanges in Bermuda virtually at par with the US dollar.

Since the mid-1960s the Bermudian economy has been heavily dependent upon international tourism to generate foreign currency, income, employment and public-sector revenue. Additional contributions have come from the presence of large NATO military stations in the country, although in 1993 the Canadian forces station was closed and the next year the US and British stations also closed. The effects might have been economically disastrous for

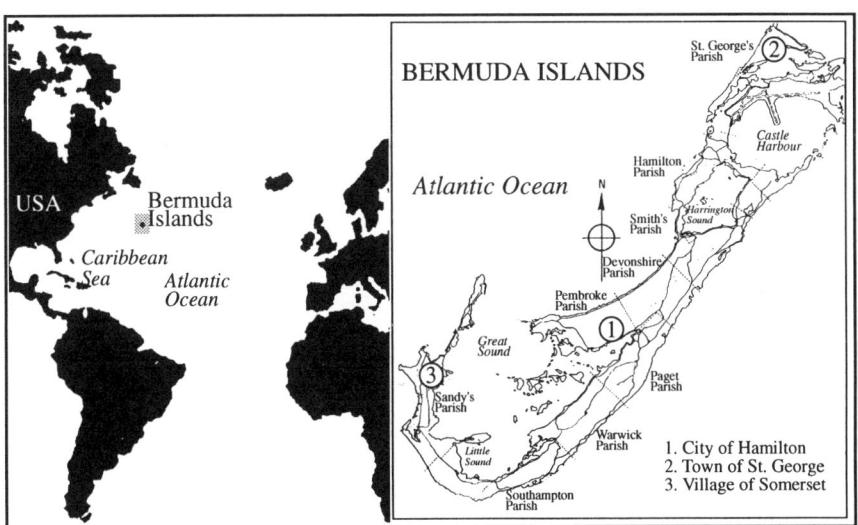

Figure 10.1 Map of Bermuda and its location

Bermuda, but fortunately the country had also developed into a major offshore financial business centre.

Indeed, international company business now makes a very substantial contribution to the Bermudian economy (see Table 10.1). Provisional estimates for 1993 and 1994 indicate that the contribution of international company businesses now exceeds that of tourism except in terms of employment generation.

Table 10.1   Comparative impacts on the Bermudian economy (1992)

Economic activity	Nature of contribution			
	Exports ($m)	Incomes ($m)	Public-sector Revenue ($m)	Employment (jobs)
International companies	358.8	509.1	124.5	4,817
US naval air station	47.1	32.0	11.9	380
Canadian forces station	3.4	3.9	1.1	46
British forces station	2.6	3.4	1.1	58
Tourism	441.5	550.0	137.2	9,550
Other exports	187.3	223.7	49.6	2,134
Total	1,040.7	1,322.1	325.4	16,986

Notes:
1. To avoid double-counting, the contribution made by the banks and BIBA companies is included with the 'international companies' where relevant (over $131 million of the $240.4 million of foreign currently earned by these Bermudian companies in 1992 was professional and bank fees paid by international businesses), and the remainder of their export activities (excluding their overseas subsidiaries and associates) are included with 'other exports'.
2. To avoid double-counting in this table, visitors to the international businesses, the US naval air station, the Canadian forces station and the British naval base are included in the 'tourism' figures and not in the other sectors.
3. The tourism income figures shown in the second column of this table include $50 million of income earned by non-Bermudians employed in those establishments which received tourist spending. The international business income figure in the same column includes $33 million of income earned by non-Bermudians in the international companies.
4. The employment figures include only direct and indirect jobs. The additional number of jobs affected by the induced effects of these export activities was 13,820.

The Nature of International Companies in Bermuda

The first offshore company was established in Bermuda during 1935 since when international businesses have increased rapidly in number. By the end of December 1993, 7,578 international companies were registered in Bermuda

(Table 10.2). Such companies can operate in Bermuda as 'exempted' companies, 'permit' companies, 'exempted' partnerships or trusts.

Table 10.2 Number of companies on register in Bermuda at the end of each year

Year	Local companies	International companies			
		Exempted	Exempted partnership	Non-resident	Total
1973	744	2,212	33	298	2,543
1974	756	2,488	30	370	2,888
1975	756	2,723	34	521	3,278
1976	796	3,038	35	546	3,619
1977	806	3,465	34	591	4,090
1978	849	3,886	38	553	4,477
1979	823	4,128	35	518	4,681
1980	884	4,519	33	569	5,121
1981	958	4,841	38	581	5,460
1982	1,002	4,959	42	606	5,607
1983	1,063	5,162	47	619	5,828
1984	1,096	5,250	50	611	5,911
1985	1,178	5,376	57	593	6,026
1986	1,238	5,531	71	545	6,147
1987	1,328	5,689	82	525	6,296
1988	1,456	5,871	92	488	6,451
1989	1,574	6,173	99	504	6,776
1990	1,653	6,389	126	487	7,002
1991	1,713	6,510	135	444	7,089
1992	1,791	6,698	155	418	7,271
1993	1,891	6,954	183	441	7,578

Source: Data supplied by Registrar of Companies
Note: The figures are net of those companies in process of liquidation.

'Exempted' companies

'Exempted' companies are called such because they are exempted from certain local ownership requirements. Incorporation is achieved by registration with the Registrar of Companies and/or by a private Act of the legislature. Exempted companies can be formed for a variety of purposes except banking, for which permission is not given.

Among the principal forms of exempted company in Bermuda are investment holding companies, trading companies, insurance companies and mutual fund companies. *Investment holding companies* can hold property or securities of any nature except shares of a local Bermuda company. There are no restrictions on the form of stock or debentures which such companies can issue except that bearer shares are prohibited. *Trading companies* can be established to carry out a wide range of trading activities outside Bermuda. Among the principal existing types of companies are sales, finance, administration and management companies. *Insurance companies* form a large component of exempted company business in Bermuda. These companies operate under the Insurance Act 1978 and related regulations which deal with matters such as the ratio of premiums to statutory capital and surplus, minimum liquidity ratios and solvency margins, and the appointment of a Principal Representative in Bermuda. *Mutual fund companies* have increased substantially in Bermuda and the country is now a major base for such funds. Before incorporation is approved, the Bermuda Monetary Authority must be satisfied with the management experience of the promoters of the mutual fund.

'Permit' companies

'Permit' companies are corporations registered in other jurisdictions which are permitted to become resident and establish a place of business in Bermuda. Thus, for example, ship-owning companies are not normally required to incorporate their businesses in the country in which the vessels are registered, but such companies can apply to the Bermuda government for permission to operate their company business from Bermuda.

'Exempted' partnerships and trusts

'Exempted' partnerships and trusts can be set up in Bermuda by individuals and companies with particular requirements. Demand for this form of organization has proved less significant than for company structures.

The Number of International Companies in Bermuda

Bermuda has sixty years of experience with offshore company development, but the principal period of growth has been the past two decades. From 2,543 registered companies in 1973, the number increased by 84 per cent to 4,681 at the end of 1979. Although the rate of increase slowed down considerably from 1981, the average annual increase in the numbers during the 1980s was 166 and so far during the 1990s has been 144. Breakdowns of the figures are given in Tables 10.2 and 10.3. From Table 10.3 it can be seen that in 1993 about 20 per cent of the companies were insurance-related; over 72 per cent were other types of exempted companies, of which the largest group was 'investment holding' (39.5 per cent), whereas exempted partnerships comprised only 2.3 per cent and permit companies only 5.6 per cent of the total.

In recent years (1987–93), the most substantial growth has been in 'mutual fund companies' (12.9 per cent), 'shipping' companies (63.3 per cent) and 'investment holding' companies (61.0 per cent). Significant growth occurred also in 'natural resource' companies (45.5 per cent) and 'other management and consultancies' (54.5 per cent).

Less than 3 per cent of these companies, however, had a physical presence in Bermuda in that they employed staff and paid land tax. According to the available records, the number of such companies in December 1993 was

Table 10.3 Number of international businesses on the register at the end of December, 1987–93

Type of business	Number of businesses						
	1987	1988	1989	1990	1991	1992	1993
Exempted companies							
Insurance	1,232	1,321	1,310	1,302	1,310	1,307	1,315
Mutual funds	130	166	198	211	213	235	298
Public finance	49	61	55	47	48	46	60
Investment holding	1,938	2,244	2,464	2,603	2,635	2,779	3,121
Commercial trading	467	506	515	493	477	437	470
Insurance broker/ manager	246	268	257	254	244	229	235
Other managers/ consultants	341	378	388	411	435	461	527
Shipping	411	491	508	550	561	569	671
Natural resources	167	183	217	229	241	257	243
Others, n.e.c.	965	544	575	602	648	707	329
1. Exempted companies total	5,946	6,162	6,487	6,702	6,812	7,027	7,269
2. Exempted partnerships	82	92	99	126	135	155	183
3. Permit companies	525	488	504	487	444	418	441
Total	6,553	6,742	7,090	7,315	7,391	7,600	7,893

Source: Data supplied by the Registrar of Companies
Notes:
1. These figures include companies in the process of liquidation.
2. Exempted insurance includes active, in run-off and dormant insurers.
3. n.e.c. = not elsewhere classified.

Table 10.4 International company businesses with a physical presence in Bermuda, 1987–93

Category group	Category	Number of companies						
		1987	1988	1989	1990	1991	1992	1993
960	Insurance – risk-takers	59	56	50	42	42	42	41
961	Insurance – brokers	19	19	21	16	18	17	15
962	Insurance – managers	35	37	39	41	39	35	36
963	Insurance – other n.e.c.	7	9	9	10	9	10	11
	Insurance – total	120	121	119	109	108	104	103
964	Shipping	13	16	16	12	10	10	13
965	Trading	37	35	33	27	27	33	33
966	Finance	32	33	25	23	27	30	30
967	Communications	5	3	4	4	1	1	1
968	Mixed operations	20	17	19	20	15	18	21
969	N.e.c.	25	36	55	68	39	39	41
	Non-insurance – total	132	140	152	154	119	131	139
	Total	252	261	271	263	227	235	242

Source: Data supplied by the Manpower Survey Unit
Note: The figures refer to the situation in a base-period week in late July/early August. (n.e.c. = not elsewhere classified.)

approximately 242. A breakdown showing the types of these companies with a physical presence in 1993 is given in Table 10.4. Of these, 42.6 per cent were insurance-related businesses.

Nature of the Impact Made by International Companies on the Economy of Bermuda

The amount of money spent in Bermuda by international companies is shown in Table 10.5. There are four principal types of expenditure: (a) Bermudian businesses receive revenue from supplying goods and services to the international companies; (b) other Bermudian companies receive professional fees for supplying legal, accounting, managerial and banking services to the international companies; (c) Bermudian households receive income as wages and salaries; and (d) the public sector receives revenue in the form of taxes, fees, duties and licences.

The recipients of this expenditure – Bermudian companies, households and the public sector – themselves respend most of the money, thereby setting in motion further rounds of economic activity within Bermuda. The money continues to flow through the economy, at each stage generating additional business revenue, income, employment and public-sector revenue, but the effects diminish progressively as money 'leaks' out of the system to purchase imports and into savings. This generating effect is known as the multiplier.

Direct Contribution Made by the Trading Activities of International Companies

From Table 10.5 it can be seen that during 1993 international companies spent $391.7 million in Bermuda. Of this sum, $208.1 million (53.1 per cent) was spent by insurance companies and $183.6 million (46.9 per cent) by other companies – a very substantial contribution to the invisible-export earnings of Bermuda. Between 70 and 80 per cent of this expenditure was contributed by the relatively small number of international companies resident in Bermuda. The principal contributions made by the large number of non-resident companies were in the form of professional fees and bank fees (c. $72 million) and payments to the public sector ($28 million).

Of the $391.7 million spent by international companies in Bermuda, $32.7 million was paid to the public sector (excluding pension deductions) and $359 million to the private sector. The principal payments to the private sector in 1993 were $143 million in wages, salaries and benefits to employees (including non-Bermudians resident in Bermuda); $135.7 million in professional fees and bank fees; $16.1 million in office rentals; and $13.2 million

Table 10.5   Direct trading activities of 'exempted' and 'permit' company businesses in Bermuda, 1993 (B$ millions)

Item		Type of company		All
		Insurance-related	Non-insurance-related	
1	Office rentals	10.5	5.6	16.1
2	Utilities and services	4.8	3.4	8.2
3	Land tax	0.4	0.3	0.7
4	Other payments to government	10.2	21.8	32.0
5	Salaries and benefits	103.0	40.0	143.0
6	Professional and bank fees	44.5	91.1	135.7
7	Communications	7.1	6.1	13.2
8	Scholarships	2.2	1.3	3.5
9	Expenditure on business visitors	4.1	2.1	6.2
10	Purchase of goods	4.1	2.0	6.1
11	Other contributions	4.9	4.3	9.2
12	Capital expenditure	12.2	5.6	17.8
	Total	208.1	183.6	391.7

Note: The data refer to expenditure by the businesses in Bermuda.

in communications. The remaining $51 million included $17.8 million of capital expenditure.

The public sector received $32.7 million plus an additional $4.3 million in pension contributions (included under wages, salaries and benefits in Table 10.5). Table 10.6 provides a detailed breakdown of this public-sector revenue. In this table, pension deductions are shown as a separate item, whereas in

Table 10.6   Direct revenue to the public sector from international company businesses, 1993 (B$ millions)

Item	Type of company		All
	Insurance-related	Non-insurance-related	
Filing, application and registration fees	0.4	1.9	2.3
Annual taxes	4.2	17.4	21.6
Land tax	0.4	0.3	0.7
	5.0	19.6	24.6
Hospital levy	5.6	2.5	8 .1
Pension deductions	3.0	1.3	4.3
	8.6	3.8	12.4
Totals	13.6	23.4	37.0

Sources: Calculated from data supplied by the Tax Commissioner, Accountant General, Registrar of Companies and Department of Social Insurance

other tables they are incorporated in wages and salaries. The main items in 1993 were annual taxes $21.6 million; filing, application and registration fees $2.3 million; land tax $0.7 million; and hospital levy $8.1 million.

Apart from their monetary contribution to the Bermudian economy, international companies employ a large number of Bermudian workers. From Table 10.7 it can be seen that in 1993, community, social and personal services (economic activity 9) accounted for a third of Bermuda's employment, of which international companies provided 2,056 jobs (6.15 per cent of total employment in Bermuda and almost 5 per cent of all Bermudian nationals' employment). These figures, however, relate only to the period of the Manpower Survey (1–7 August 1993). Data from the Office of the Tax Commissioner show that during 1993, employment in International Companies ranged from 1,951 in the first quarter to 1,935 in the second quarter, 1,933 in the third and 1,866 in the fourth, giving an average for the year of 1,921.

Further details of employment in the international companies are provided in Tables 10.8–10.10. The number of companies of each type with a physical presence in Bermuda and the size of the labour force plus vacancies are shown in Table 10.8. A more detailed breakdown of the labour force by nationality and sex is given in Table 10.9. The two tables together show that, although the non-insurance companies with a physical presence in Bermuda outnumbered insurance companies (139 as against 103), they employed

Table 10.7 Summary analysis of employment by nationality and by economic activity (number of workers, Bermuda, 1993)

Economic activity		Nationality			Non-Bermudian as % of total in each economic activity
		Bermudian	Non-Bermudian	Total	
1	Agriculture and fishing	251	135	386	35.0
2	Quarrying	97	8	105	7.6
3	Manufacturing/servicing	859	182	1,041	17.5
4	Electricity, gas, water	480	33	513	6.4
5	Construction	1,507	240	1,747	13.7
6	Wholesale/retail trade and restaurants/hotels	8,566	2,175	10,741	20.2
7	Transport, storage and communications	2,214	115	2,329	4.9
8	Finance, insurance, real estate and business services	4,206	1,197	5,403	22.2
9	Community, social and personal services	8,306	2,856	11,162	25.6
	Total	26,486	6,941	33,427	20.8

Source: Adapted from Ministry of Labour and Home Affairs, *Report of the Manpower Survey, 1993*, Bermuda Government, December 1993

Table 10.8 International company businesses with a physical presence in Bermuda, 1993, by category group and number of employees

Category group	Category	Number of companies	Size of labour force		
			Employees	Vacancies	Total
960	Insurance – risk-takers	41	562	8	570
961	Insurance – brokers	15	153	7	160
962	Insurance – managers	36	688	10	698
963	Insurance – others n.e.c.	11	27	1	28
	Insurance – total	103	1,430	26	1,456
964	Shipping	13	59	1	60
965	Trading	33	186	2	188
966	Finance	30	104	2	106
967	Communications	1	2	0	2
968	Mixed operations	21	131	1	132
969	n.e.c.	41	144	4	148
	Non-insurance – total	139	626	10	636
	Total	242	2,056	36	2,092

Source: Data supplied by the Manpower Survey Unit
Note: The figures refer to the situation in the period 1–7 August 1993. (n.e.c. = not elsewhere classified.)

Table 10.9 Analysis of labour force in international company business by status and sex, 1993

Category group	Category	Bermudian			Non-Bermudian			All			Vacancies	Total
		Male	Female	Total	Male	Female	Total	Male	Female	Total		
960	Insurance – risk-takers	93	300	393	116	53	169	209	353	562	8	570
961	Insurance – brokers	28	70	98	32	23	55	60	93	153	7	160
962	Insurance – managers	77	365	442	159	87	246	236	452	688	10	698
963	Insurance – others n.e.c.	2	7	9	14	4	18	16	11	27	1	28
	Insurance – total	200	742	942	321	167	488	521	909	1,430	26	1,456
964	Shipping	3	34	37	19	3	22	22	37	59	1	60
965	Trading	35	87	122	45	19	64	80	106	186	2	188
966	Finance	20	49	69	23	12	35	43	61	104	2	106
967	Communications	2	0	2	0	0	0	2	0	2	0	2
968	Mixed operations	10	53	63	47	21	68	57	74	131	1	132
969	N.e.c.	21	57	78	39	27	66	60	84	144	4	148
	Non-insurance – total	91	280	371	173	82	255	264	362	626	10	636
	Total	291	1,022	1,313	494	249	743	785	1,271	2,056	36	2,092

Source: Data supplied by the Manpower Survey Unit
Note: The figures refer to the situation in the period 1–7 August 1993. (n.e.c. = not elsewhere classified.)

Table 10.10   Occupations of employees in ICD member companies, August 1993

Occupation	Insurance-related			Non-insurance related			All		
	Bermudian	Non-Bermudian	Total	Bermudian	Non-Bermudian	Total	Bermudian	Non-Bermudian	Total
Managers	60	147	207	37.5	72	109.5	97.5	219	316.5
Accountants	62	101	163	28.5	31	59.5	90.5	132	222.5
Bookkeepers	115	3	118	39.5	2	41.5	154.5	5	159.5
Executive secretaries	55	23	78	39.5	9	48.5	94.5	32	126.5
Secretaries and typists	64	6	70	23	5	28	87	11	98.0
Clerks	79	4	83	33	2	35	112	6	118.0
Underwriting staff	103	31	134	1	0	1	104	31	135.0
Computer/ data-processing staff	40	18	58	8	8	16	48	26	74.0
Others	48	5	53	9.3	2	11.3	57.3	7	64.3
Total	626	338	964	219.3	131	350.3	845.3	469	1,314.3

Source: Data supplied by ICD, Bermuda Chamber of Commerce

fewer workers – 626 (30.4 per cent) as against 1,430 (69.6 per cent). Both types of companies employed far more Bermudian than non-Bermudian workers (for insurance-related companies 65.9 and non-insurance companies 59.3 per cent, giving a weighted average of 63.9 per cent). Female workers accounted for 61.8 per cent of the total (63.6 for insurance companies and 57.8 per cent for non-insurance). In respect of Bermudian workers only, 22.2 per cent of the workforce was male compared with 66.5 per cent for non-Bermudians. The reason can be deduced from Table 10.10, which shows the occupational breakdown of the labour force in the International Companies Division (ICD) of the Bermuda Chamber of Commerce survey: only 22.2 per cent of the Bermudian workers were employed as managers and accountants compared with 74.8 per cent in the case of non-Bermudian workers, whereas 66 per cent of the Bermudian workers were employed as bookkeepers, secretaries, typists, clerks, etc., compared with 17.7 per cent in the case of non-Bermudians.

The *direct* contributions made by the trading activities of international companies to the economy of Bermuda in 1993, therefore, was:

- $391.7 million to invisible exports;
- $208.2 million to business activity among Bermudian companies, including utilities (but excluding pension deductions);
- $37.0 million to public-sector revenue, including pension contributions of $4.3 million;

- $143 million in incomes (salaries and benefits) including those paid non-Bermudians, plus $3.5 million in scholarships;
- 1,866–2.056 jobs.

The Secondary Contribution Made as a Result of the Trading Activities of International Companies

The money spent in Bermuda by the international companies forms revenue paid to Bermudian companies and the public sector, and income to households. A large portion of this revenue and income is respent in Bermuda, creating secondary flows of money within the economy, thereby generating additional revenue, income and employment. The process continues through further rounds of expenditure, but diminishing in amount as money leaks abroad to purchase imports and into savings.

During 1993, $135.7 million of the $391.7 million spent by international companies was received as professional and bank fees by Bermudian businesses, mainly members of BIBA (Bermuda International Business Association).

The BIBA companies

The BIBA members are individuals representing companies dealing with banking, accountancy and finance, law, management and insurance management. Many of these companies, especially the larger ones, carry out several functions in addition to their principal role. The banks, for example, provide management, accounting, financial and legal services in addition to banking, while the largest law firms provide accounting, financial and management services in addition to legal ones. It is inappropriate, therefore, to regard individual firms as single entities performing specific narrowly defined functions.

Some of the BIBA members derive the whole or a major part of their revenue from the international companies, whereas others obtain significant revenue from other sources – both local and international.

The banks

The three banks have total assets in excess of $10 billion and employ some 2,155 people in Bermuda alone. During 1993 the Bermuda-based operations of the three banks earned a total of $146.6 million net foreign exchange from all sources. This figure excludes the receipts earned abroad by the subsidiaries and associates of these banks. Of these net foreign exchange earnings, $57.7 million (39.4 per cent) was in the form of professional and bank fees paid by the international companies. These fees included investment fees, foreign exchange, interest margin and other fee revenues.

Local-currency sources earned a further $57.2 million. Total employment in Bermuda in the three banks in 1993 averaged 2,155 of whom 1,746 (81.0 per cent) were Bermudians. Only part of this employment, however (less than 40 per cent), was related to their work for international companies.

Other BIBA member companies, providing accounting, financial, legal, management and insurance management services

During 1993, these other companies received $82.5 million in foreign currency earnings of which $78 million was received from the international companies. In addition they received $22.8 million from local sources. Total employment in these companies was 1,162, of whom 746 (64.2 per cent) were

Table 10.11   Number of employees in ICD member companies, 1977–93

Year	(Month)	Bermudian		Non-Bermudian	Total	Response
1977	(December)	—		—	994	85 companies
1978	(June)	677	(64.0%)	303	980	
1978	(December)	—		—	1,189	87 companies
1978	(June)	802	(64.6%)	440	1,242	
1980	(December)	—		—	1,538	113 companies
1981	(April)	1,012	(63.9%)	572	1,584	
1981	(December)	—		—	1,770	133 companies
1982	(April)	1,169	(63.9%)	600	1,769	
1982	(December)	—		—	1,759	127 companies
1983	(May)	1,090	(63.2%)	634	1,724	
1983	(December)	—		—	1,718	122 companies
1984	(May)	1,134	(65.3%)	603	1,737	
1984	(December)	—		—	1,528	123 companies
1985	(April)	988	(64.4%)	546	1,534	
1985	(December)	—		—	1,316	101 companies
1986	(April)	860	(64.7%)	470	1,330	
1987	(April)	898	(66.6%)	450	1,348	93 companies
1987	(August)	851	(66.2%)	435	1,286	
1988	(August)	897	(65.6%)	471	1,368	103 companies
1989	(August)	966	(64.7%)	526	1,492	107 companies
1990	(August)	946	(68.7%)	432	1,378	93 companies
1991	(August)	918	(68.5%)	422	1,340	101 companies
1992	(August)	924	(67.5%)	445	1,369	95 companies
1993	(August)	845	(64.3%)	469	1,314	95 companies

Source: Data supplied by Bermuda Chamber of Commerce
Note: The figures in parentheses show Bermudian employees as a percentage of the total.

Table 10.12 Direct contribution made by BIBA member companies to the economy of Bermuda, 1993

Item		Banks	Other companies	All
Receipts				
1.	Net foreign exchange receipts from all sources	146.6	82.5	229.1
	Of which receipts from 'exempted' and 'permit' companies	57.7	78.0	135.7
2.	Net receipts from domestic sources	57.2	22.8	80.0
	Total ($m)	203.8	105.3	309.1
Expenditure in Bermuda				
3.	Wages and salaries	90.6	43.6	134.2
4.	Other expenditure	86.8	30.6	117.4
	Total ($m)	177.4	74.2	251.6
Employment				
5.	Non-Bermudian employees	409	416	825
6.	Bermudian employees	1,746	746	2,492
	Total	2,155	1,162	3,317

Source: Archer, B.H., BIBA survey, 1993. The figures are the aggregated responses (adjusted for non-respondents) to the BIBA survey
Note: Net foreign exchange receipts *exclude* receipts earned from the operations of overseas subsidiaries and associates.

Table 10.13 Overall (direct plus secondary) contribution made by the trading activities of international companies to the economy of Bermuda, 1993 (B$ millions)

Item	Type of company		
	Insurance related	*Non-insurance-related*	*All*
Income			
Direct[a]	103.0	40.0	143.0
Indirect	56.7	86.3	143.0
Induced	150.6	119.1	269.7
Total	310.3	245.4	555.7
Public-sector revenue			
Direct	10.4	22.1	32.5
Indirect	11.0	12.9	23.9
Induced	44.5	35.2	79.7
Total	65.9	70.2	136.1
Balance of payments			
Exports – total	208.1	183.6	391.7
Imports – Direct and indirect	24.8	29.1	53.9
Induced	151.3	119.5	270.8
Net impact	32.0	35.0	67.0

Note:
[a]Direct income includes salaries paid to non-Bermudians.

Bermudians. It is estimated that just under half of this employment is dependent upon international company business.

The overall contribution of the banks and other BIBA member companies to the economy of Bermuda is shown in Table 10.11. During 1993 they paid out $134.2 million in wages, salaries and benefits and an additional $117.4 million in other expenditure (including capital expenditure) in Bermuda. These expenditures generated further tertiary activity: additional income, employment and public-sector revenue. These effects are included where relevant in the indirect and induced impact made by the international companies; see Tables 10.12 and 10.13.

Other expenditure in Bermuda by the international companies

Apart from the $135.7 million paid as professional and bank fees, the international companies spent an additional $256 million in Bermuda.

Of this sum, $143 million was paid as wages, salaries and benefits, $32.7 million to the public sector, $16.1 million as office rentals, $13.2 million on communications, $17.8 million on capital expenditure and the remaining $33.2 million on utilities and services, the purchase of goods, on business visitors, scholarships and other contributions; see Table 10.5. This expenditure comprised revenue to Bermudian companies and the public sector and income paid to local households. As this money was respent it created further rounds of economic activity within Bermuda.

The total amount of income, public-sector revenue and employment and the effect on the balance of payments is shown in Tables 10.12 and 10.13. The indirect effects shown in these tables are those created by the trading which takes place between the various sectors of the Bermudian economy and the induced effects are the additional effects generated by the respending by Bermudian residents of the incomes earned as a direct or secondary consequence of international company activity.

These tables include the impact made by BIBA member companies as a consequence of revenue received from international companies, but exclude the impact made by receipts from other overseas sources.

Contributions Made by the Expenditure of Business Visitors to the International Companies and to BIBA Firms

It is estimated that during 1993, over 13,000 business travellers visited the international companies and BIBA members, together with approximately 11,000 friends and relatives. Many of these visitors received hospitality in the form of free accommodation, meals, etc. from the ICD companies ($6.2 million in 1993) and BIBA companies ($1.1 million). Their remaining expenditure of $18.4 million generated some $22 million of income, $5 million of public-sector revenue, occasioned over $16 million of imports and helped to support over 200 jobs.

Summary

The data in this summary include the impact made by business visitors to the international companies and BIBA members.

Balance of payments

Expenditure in Bermuda by the international companies and their business visitors accounted for $410.1 million in 1993. This was partially counter-balanced by $56.6 million worth of imports needed to service these companies and their visitors, and a further $284 million worth of imports to satisfy the demand from Bermudian households. The overall net impact on the balance of payments, therefore, was +$69.5 million.

Public-sector revenue

The international companies and their visitors contributed $141.1 million to the public-sector purse (including all secondary effects).

Income

International companies and their visitors generated $577.7 million of income in Bermuda of which $148 million was direct income.

Employment

International companies and their visitors maintained almost 5,000 jobs in the country and severely influenced a further 5,500.

Conclusions

Quite deliberately, no attempt has been made in this chapter to generalize the findings of the study to other islands or small states. Although the model itself is suitable for this purpose and has been employed for a variety of purposes in many countries, each country's economy is unique and the economic impact made by international companies in any particular country depends upon the extent to which the activities of these companies are integrated into the economy; that is, the extent to which the economy is able to supply the international companies with goods and services, especially professional services.

Much greater detail about the studies undertaken in Bermuda can be gleaned from two recent reports (Archer, 1994a, b).

References

Archer, B.H. (1973) *Tourism Multipliers: the State of the Art*. Cardiff: University of Wales Press.
Archer, B.H. (1994a) *The Bermudian Economy: the Impact of Export Earnings in 1992*. Ministry of Finance, Government of Bermuda.
Archer, B.H. (1994b) *International Companies 1993*. Ministry of Finance, Government of Bermuda.

11

The Liberalization of Interest Rates in Malta: Policies and Strategic Options

Paul V. Azzopardi and Lino Briguglio

Until 1992, interest rates on lending and borrowing by banks and financial institutions in Malta (see Figure 11.1) were rigidly controlled by the government and established by the Minister of Finance. Interest rates were kept relatively low, and in some years the real rate of interest on one-year fixed deposits was actually negative.

Interest rate control was just one of a range of economic restrictions affecting prices, profit margins, foreign exchange transactions and foreign trade. Foreign exchange control in Malta was particularly severe, with heavy fines and even imprisonment threatened to offenders. This notwithstanding, the prevailing low interest rate regimes gave rise to very large leakages (Tarling and Rhodes, 1989), in pursuit of higher rates of returns and tax evasion.

The wave of deregulation and liberalization which has hit many other countries has also affected Malta, and a process of liberalization has been ushered in during the past few years. The fact that the Maltese government applied to join the European Union in 1990 brought a degree of urgency to the process of liberalization, and the process has now reached an advanced stage.

This chapter discusses the impact of interest rate liberalization and argues in favour of a gradual approach so as to establish the necessary preconditions for an orderly transition. Following this introduction, the interest rate regime in Malta is briefly described. There follow a discussion of the pros and cons of interest rate liberalization, and a consideration of the implications of interest rate liberalization in Malta. The strategic options for Malta are then discussed.

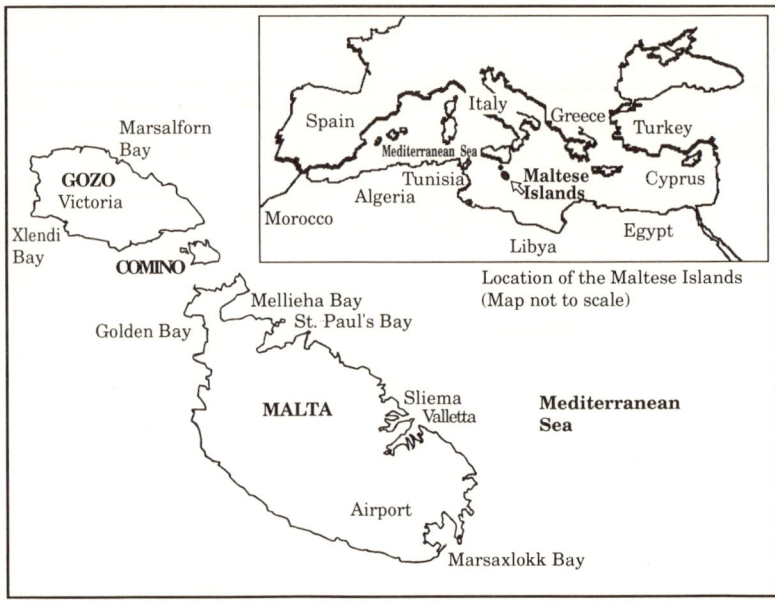

Figure 11.1 The Maltese islands and their location in the Mediterranean Sea

Interest Rate Regime in Malta

Monetary activity is regulated in terms of the Central Bank of Malta Act 1967 (as amended in 1994), the Banking Act 1994, and the Financial Institutions Act 1994 and the Exchange Control Act 1972. A broad spectrum of legislation and legislative amendments was also recently enacted to foster the development and proper functioning of the financial services sector in Malta.

Up to 1993, lending rates were subject to a ceiling set by the Minister of Finance, while deposit rates were similarly constrained by a limit imposed by the Central Bank. Preferential lending rates were also set for so-called priority sectors. The banking sector, which up to that year was owned by the public sector through majority shareholding, could not operate competitively in the circumstances, and the limits were taken as the ongoing market rates. This occurred within the context of a scantily developed financial sector, where demand for money was passively met, reflecting mainly fiscal and balance of payment balances. In view of the fixed interest rate regime, restrictions on foreign exchange transactions were also imposed, to sustain the external value of the Maltese lira.

Interest rates on bank deposits, loans and advances were set by law, which gave the Minister of Finance complete control over maximum rates.

The first authoritative pronouncement in favour of interest rate liberalization in Malta was put forward by the Governor of the Central Bank of

Malta (Galdes, 1989, 1990a), who on two occasions argued in favour of a more efficient financial market, which, according to him, necessitated a process of deregulation. This, however, necessitated major reforms in the institutional set-up of the financial sector.

Subsequently, there were further written contributions on this issue (see Azzopardi, 1991; Consiglio, 1991; Cordina, 1992; Cutajar, 1992; Guillaumier, 1991; Dean, 1993; Vassallo, 1994a, 1994b and 1995), generally favouring liberalization. There were many statements by Mr John Dalli, Maltese Minister of Finance (see Malta 1991), acknowledging the advantages of interest rate liberalization but cautioning against a hurried approach in dismantling controls.

Another important statement in this regard was made by Joseph Tabone in 1991 as Chairman of Bank of Valletta, one of the two leading banks in Malta. Mr Tabone argued in favour of interest rate liberalization. Referring to the distortions arising from a government-imposed structure, he commented on the need by commercial banks in Malta to demand very high collateral and to allocate funds in low-risk economic activity, since the current interest rates could not adequately cover risk premiums (see Tabone, 1991).

The initial steps towards a deregulated financial system took place in the late 1980s and the early 1990s, when the government reduced its shareholding in the banks, the banking supervisory structure was consolidated and the reserve requirements were activated, while the Maltese commercial banks' capital structure was solidified.

A major development regarding interest rate policy occurred in 1993. This was the linking of the maximum interest rates charged by commercial banks on loans and advances to the Central Bank of Malta's discount rate (see Central Bank of Malta, *Annual Report*, 1992, p. 67). The preferential lending rate scheme was simplified. Up to that year, the number of borrowing categories, as determined by the Minister of Finance, amounted to eight, with different permitted maximum interest rates for each category. In 1993, the number of categories was reduced to four, namely inter-bank lending for housing (1 percentage point above Central Bank discount rate), lending for residential housing (1.5 per cent above discount rate), lending to priority economic sectors and lending to so-called 'other lending' (3 per cent above the discount rate).

In 1994, interest rates were further liberalized, and commercial banks were allowed to establish their own interest rates on all types of deposit accounts, provided that a minimum of 3 per cent was maintained on savings and term deposits for any maturity up to a maximum of 5 years.

Furthermore, the preferential lending scheme was abolished, with a rate 3 percentage points above the Central Bank discount rate being imposed for all lending (except for housing-related credit, where the maximum was 1.5 percentage points over the discount rate for the first Lm 15,000) and for funds lent for speculative purposes (where the limit was raised to 3.5 percentage points over the discount rate).

Thus by 1994, the Central Bank was given almost complete authority over interest rate policy. However, it was still not clear whether the margin over the discount rate at which maximum lending rates were set was a prerogative of the Central Bank or the Minister of Finance. In anticipation of the total removal of the controls on interest rates, the Central Bank sought methods and avenues by which it could influence interest rates through market-oriented means. Thus, the Central Bank further enhanced its ability to give out signals to and influence the money market by bidding directly for treasury bills in competition with commercial banks and other financial institutions.

Moreover, the possibility of carrying out open-market operations had already improved with the opening up of the Stock Exchange in 1992. The intervention of the Central Bank stockbroker on the Stock Exchange was aimed at fostering activity in the market and at the same time helping to promote stable conditions in line with monetary policy.

In 1994, the Central Bank introduced an important instrument into the Maltese monetary system, namely repurchase agreements (called seven-day repos) whereby the Central Bank provides short-term liquidity to the banks and, when required, absorbs liquidity from the banks. The system works like an auction, with the banks bidding within an interest rate band determined by the Central Bank, thereby permitting money market interest rates to respond, within limits, to market conditions. The seven-day repo rate is considered by the Central Bank as a more indicative money market rate than the discount rate, and the latter has not been used for this purpose as from 1995, although it still retains a function for legal reasons of stipulating a ceiling. In this way the Central Bank injects or absorbs funds temporarily into the banking system, with the difference between buying and selling price representing the interest rate charged.

In 1994 there were also changes in exchange controls. Resident companies and persons were allowed to invest up to Lm 50,000 a year outside Malta in going concerns, and the normal investment allowance was increased to Lm 5,000 annually. However, exchange controls remained in place.

In 1995, bank lending rates were liberalized virtually completely, in the sense that the ceiling was increased from 3 percentage points above the discount rate to 10 percentage points, although legal requirements precluded the Central Bank from completely freeing interest rates, and a ceiling had to remain in operation. However, the 10 percentage point margin to all intents and purposes rendered the ceiling ineffective. The notable exception is borrowing for residential housing, which was retained at 1.5 percentage points over the discount rate.

The ceiling on interest rates on non-Maltese lira funds was also increased from 5 to 10 percentage points over the London Interbank Offered rate (LIBOR).

To conclude, therefore, interest rates in Malta are to a large extent liberalized, but the legal framework permitting the authorities to impose ceilings is retained, although rendered ineffective by the high ceilings chosen. The monetary authorities in Malta would seem to favour a system which retains the option of maximum rates, so as to permit legal intervention if the mechanism gets out of hand. The changes just described took place over a period of just three years, but they were introduced in phased stages, presumably in line with the Maltese government's objective of preparing Malta for accession to the European Union, as indicated by the Malta government in response to the EU *avis* on Malta's entry into the Union (Malta, Government of, 1991–5).

Interest Rates and the Macro-economy

As explained, the interest rate regime in Malta prior to 1993 was characterized by direct government controls. This section discusses the pros and cons associated with such a regime.

Advantages of direct control

Reducing uncertainty

Direct control is an effective method of keeping interest rates at a low level, thereby encouraging investment. By fixing interest rates the uncertainty concomitant with fluctuating rates is removed. The cost of borrowed funds, except for charges in respect of banking services, can be accurately forecast provided tax rates remain constant. In fact, avoiding uncertainty by keeping rates fixed is often more important than keeping them low. Low interest rates and the abatement of uncertainty are considered to be prime incentives for the investment in industrial projects that is so necessary for development. Foreign and local equity capital is attracted in the knowledge that the bulk of finance will be forthcoming from the banking sector.

Allocating funds to priority areas

Direct control is an effective method of allocating finance to sectors and industries which are considered priority areas. Under a controlled regime, funds are channelled towards designated sectors which are thought to be of national importance. Low interest rates are often supported by government guarantees to financial institutions lending to these sectors. Such institutions would otherwise not willingly lend at reduced rates of interest to sectors which, although important, often carry a high risk of default.

Controlling inflation

Keeping interest rates fixed is conducive to low inflation rates, since interest rate-induced inflationary pressures are an important component of the cost of production and consumption.

Controlling costs of public debt

Direct controls, leading to low interest rates, would keep the cost of servicing public debt low. This would in turn have repercussions on government's budget and fiscal policies.

Countervailing oligopolistic structures

In a small economy such as Malta's, the banking sector is bound to be oligopolistic. In such circumstances, interest rate control may be better then interest rate determination by an oligopolistic banking industry operating in thin financial markets. Indeed, developed financial markets and competition among banks and financial institutions are considered to be essential pre-requisites if interest rate liberalization, especially if rapidly implemented, is not to have serious side-effects.

Disadvantages of direct control

Low interest rates may have a negative impact on investment. One argument in favour of direct control is that such controls can keep interest rates low to stimulate investment. The relation between low interest rates and investment has been questioned by a number of authors, most notably McKinnon (1973) and Shaw (1973). They have argued that if interest rates are established at a level below equilibrium, demand for investment will increase, but many potential investors will be frustrated owing to lack of savings. As is well known, an excess demand regime means that the transacted amount will be constrained by the short side of the market, which in this case would represent savings.

The inefficiency in the channelling of savings funds to productive applications within a fixed interest rate regime can be illustrated by what happened in Malta in the second half of the 1980s. In 1987, the commercial banks had accumulated Lm 120 million (about 25 per cent of total assets) in the form of deposits with the Central Bank, which are low-interest but risk-free assets. This excess liquidity could have been eliminated by a downward adjustment in interest rates (including those offered by the Central Bank on commercial banks' deposits) to stimulate lending for consumption and investment purposes. After 1987, hefty government borrowing depleted excess bank reserves with the Central Bank to the extent that the private sector has to an extent been crowded out from the capital market. Market pressures would have brought about an upward adjustment in interest rate, which in the existing controlled regime has not been forthcoming.

Problems of discretionary allocation of funds

The inability to find investment funds in a low interest rate regime may lead to rationing by financial intermediaries or discretionary allocation by the government. Funds may be assigned towards investment with a poor rate of return but with good collateral, or towards investment which yields political advantages.

Discretionary allocation of funds on the basis of differentiated interest rates may also be counter-productive, mostly because funds may be rechannelled to areas which are not considered priority sectors (Khatkhate and Villanueva, 1978).

In the case of Malta, it has been argued that the system of direct controls has produced very unsatisfactory results, foremost among which is that bank loans have tended to be extended to the business sector under very stringent conditions to avoid risk, concomitant with the low rate of interest charged on advances. There was therefore a credit crunch situation in Malta, directly arising from the highly regulated system, and this gave rise to the development of an 'underground' credit mechanism – a curb market possibly reflecting market forces, where the rates of interest charged vary considerably.

Guillaumier (1991) has shown that the banking sector has been giving a very small contribution to business investment in Malta, attributing this fact to what he termed the 'collateral syndrome' arising from the rigid interest rate regime.

Demand for non-interest-bearing assets

Low interest rates may be conducive to demand for non-interest-bearing assets. If interest rates are low, especially if they are lower than the expected rate of inflation, postponing consumption by acquiring interest-bearing assets would not be a viable proposition and participants in the economy would tend to accumulate non-interest-bearing assets. These include real assets such as gold and precious stones, or currency which may be directed towards underground economic activity for tax evasion purposes.

In the case of Malta, the size of the underground economy in the late 1980s has been estimated at between 20 and 25 per cent of GDP (Briguglio, 1989). The fixed low interest rate regime was probably an important factor affecting the demand for currency, which fuels this type of economic activity. In Malta currency demand is one of the highest in the world in relative terms, averaging around 40 per cent of GNP in recent years and amounting to around US$3,000 per capita in 1995.

Capital flight abroad

Alternatively, income earners may attempt to acquire foreign financial assets provided the rate of return abroad is sufficient in the light of expected changes in the foreign exchange rate or the expected rate of inflation, and provided that there is a legal or illegal outlet for such forms of financial investment. In the case of Malta, considerable amounts of funds were invested abroad, despite the fact that the foreign exchange market is highly controlled. The magnitude of such leakages was not known, but it has been estimated that in the late 1980s around 25 per cent of tourist expenditure in Malta flowed back out of Malta through illegal channels (Tarling and Rhodes, 1989).

Ineffectiveness of controls

Other arguments against interest rate controls relate to the cost of implementing and maintaining them and to the effectiveness of such controls. Once in place, controls create a strong incentive to avoid them, contain the seed of their own destruction and yet necessitate further controls to prop them up. Sectors which for non-economic reasons are considered to warrant being given priority may acquire unjustified amounts of funds. Unofficial financial intermediaries emerge to satisfy demands not catered for by official intermediaries – a process erroneously called disintermediation. Official intermediaries, on their part, seek to increase their returns by pushing up fees and by shortening compounding intervals.

Narrow range of instruments

Another disadvantage of controlled interest rates is that revolving short-term credit provided by commercial banks becomes the cornerstone of the financial market. Little attention is devoted to providing long-term finance. Differentiated financial instruments providing investors and borrowers with a choice of risks, returns and maturities are not developed. Financial and capital markets do not emerge, and the financial structure remains primitive.

The arguments against a controlled interest rate regime are therefore formidable and have prompted many governments to embark on a process of liberalization.

Implications of Liberalization

In theory, liberalization would seem to be a better arrangement than direct government controls, principally because market forces should by themselves establish an equilibrium interest rate to clear the market. The continuous movement towards an equilibrium interest rate is the basic argument underpinning those who advocate interest rate liberalization for the purpose of efficient allocation of financial resources. In addition, a system based on the operation of market forces is cheaper to operate and works against corruption, because the system does not require a huge state apparatus to run it.

However, market adjustment may not operate as in the 'textbook' model because of three main considerations: market imperfections, undesirable effects of free markets; and difficulties related to the recognition of whether or not the market is operating properly.

Market imperfections

Oligopolistic tendencies

A factor to consider in the context of market imperfections is the oligopolistic character of the banking industry, to which reference has already been made. If interest rate control by government is not to be merely substituted by interest rate control by banks, it is necessary to foster competition in bank deposits and credit.

In a small economy, such as the Maltese one, the market cannot support more than a few domestic banking institutions. This drawback, in terms of competition, can be effectively neutralized if exchange controls are dismantled. The complete removal of exchange control, however, may not be warranted in the short run for macro-economic reasons.

In the interim, other ways of fostering competition may be sought, including, for example, the opening of branches of foreign banks. In Malta, a licence to the Midland Bank has recently been issued, enabling it to offer banking services on the island. This could push local banks to behave competitively in order to reduce the threat of foreign banks' entering the domestic market, attracted by monopoly profits.

Moral suasion by the Central Bank could also be useful in fostering competition between the Maltese banks.

Another way of fostering competition in the financial sector is through legal means. In Malta, the Fair Trading Act has recently been enacted for controlling monopolistic practices in all sectors of the economy.

Sluggish adjustment

Sluggish adjustment may also be considered as a market imperfection. If banks are left free to determine deposit and lending rates, interest rates may not respond rapidly enough to changing monetary and international developments, and the resulting sluggishness may bring about some of the adverse factors associated with interest rate control (see Leite and Sundararajan, 1990). This could take place especially in the period following deregulation, as the operators in the financial markets might require time to adjust to the new operating environment. In Malta, for example, bank retail rates were changed only in 1996, in spite of the fact that interest rates were virtually liberalized in 1995 and the Central Bank had been exerting upward pressure on money market interest rates throughout that year.

If the allocation of resources is to be efficient, interest rates must be free to respond rapidly; industry-induced frictions are as unhealthy as government-imposed regulations. On the other hand, rapid response can be secured only by sacrificing interest rate and overall macro-economic stability.

Market distortions

A third factor connected with market imperfections is the role of government borrowing, which, if excessive, may create extensive distortions. In Malta, for example, the modernization and expansion of infrastructural services are necessitating budget deficits which are being largely financed by the issue of government securities. While government paper will serve to lay the basis for a future money market, the need to issue government securities may itself push up interest rates and crowd out the private sector. A large stake in domestic credit by the public sector will therefore impair the effectiveness of interest liberalization, especially in the case of Malta, where the government still retains majority shareholding in one of the two major banks and substantial shareholding in the other.

Unequal operators in the market

Another factor giving rise to market imperfection is that financial institutions differ in size and power, and therefore do not benefit equally from the liberalization process. Some might be in a position to demand a lower margin or may lower collateral requirements, thus increasing the relative importance of project feasibility in lending decisions. Also, with higher interest rates certain borderline loans and advances may become non-performing and as interest due is capitalized they may absorb some of the increased credit capacity. The value of banks' holdings of fixed-interest securities will also fall, thus reducing the asset base. The extent of such borderline loans and advances and the ability of the banking system to withstand adversity are important considerations in this regard.

This list of market imperfections suggests that liberalization might not automatically give rise to efficient allocation of resources, as implied in textbook equilibrium solutions, where the notional and effective demand and supply curves are generally assumed to be the same. Before one embarks on a process of interest rate liberalization, it is necessary therefore first of all to understand market imperfections within the economy and try to minimize them with the aim of facilitating the workings of the market mechanism.

Undesirable effect of free markets

There are a number of factors which may lead one to conclude that even if market forces work, their effect is not necessarily desirable.

Risk aversion of banks

The first factor we consider in this regard is related to risk aversion. Let us for the sake of the argument assume that banks are allowed to raise interest rates until the market clears. Modern financial theory strongly suggests that in the presence of strict bank supervision – prudential regulation requiring banks to make adequate provision against loan losses and appropriately priced deposit insurance – banks behave as risk-averse participants and will not raise rates to the limit where the market clears (Villenueva and Mirakhor, 1990). Banks will voluntarily opt for lower rates.

The rationale for this behaviour is as follows. As the rate of interest is increased the probability of repayment falls. At high market clearing levels creditworthy borrowers would drop out of the market and borrowers with high-return (and high-risk) projects, but with low repayment probabilities, would flood in.

Beyond a certain point the banks' expected return – which is equal to the product of interest rate and repayment probability – falls and banks will not raise interest rates beyond this point. This optimal point, where banks' expected return is maximized, will be lower than the market clearing rate.

On the other side of the coin, there is the possibility that in the absence of strict bank supervision and prudential regulation interest rates may be pushed up to undesirable levels in a free market. This is especially so in the presence of free deposit insurance (such as government protection). Under such circumstances, banks will be free to raise interest rates to the limit and take high risks knowing that, if successful, they stand to reap large gains; if unsuccessful, depositors and government will foot the bill. Financial chaos will ensue.

In Malta, bank supervision and prudential regulation have been considerably strengthened in recent years (Galdes, 1990a and b; Central Bank of Malta, *Annual Report*, 1995). Deposit insurance, however, does not exist. There is only the general belief that government or the Central Bank would step in should a bank find itself in difficulties – an implicit protection system (Mas and Tally, 1990). Since such protection is being provided free of charge, a risk-seeking bank would not be penalized to the same extent had an explicit deposit insurance scheme existed. Explicit deposit insurance would give the authorities another lever of control which, complemented by effective bank supervision, would prevent banks, post-liberalization, from participating in projects at the top end of the risk spectrum. Banks exhibiting risk-seeking behaviour would have to pay higher deposit insurance premiums.

Reductions of policy options

Another effect of market liberalization which some would consider undesirable is that the market will not necessarily allocate funds to those sectors which society, or its policy-makers, consider 'merit' areas. Priority sectors which in Malta, up to 1992, benefited from low interest rates might have to be subsidized through other channels, otherwise there would be many casualties. Foreign investors, who are in the habit of making relatively small direct equity investment and gearing it up with local bank borrowing, may have to be offered compensating incentives.

In this context, it may be possible that interest rate liberalization will impose additional demands on taxpayers. If depositors are to get an enhanced compensation for their deposits via substantially positive interest rates, taxpayers will have to carry the burden previously borne by depositors. Taxation may have to rise both to carry higher subsidies and to finance the higher cost of public debt. If government attempts to give lenders special incentives to hold government debt (e.g. by paying tax-free interest), it will effectively be pushing up the interest rates which the private sector has to offer to attract funds.

A related consideration in this regard is that interest rate liberalization is likely to call for liberalization in other areas of the economy – such as foreign exchange. If the market is to be left free to allocate resources efficiently, all controls must be reviewed, be they profit margin controls, price controls, direct and indirect trade controls or foreign exchange controls. If borrowers are to face increasing interest rate costs they cannot be expected to continue operating in the same strait-jacket imposed while interest rates were both fixed and low. This situation will of course reduce the government's ability to use discretionary policies for specific targets and sections of society. This is a price society will have to pay to attain overall improvements in the allocation of resources.

Time inconsistency

An undesirable effect of liberalization is that unless the process is properly administered, it may introduce expectations which distort the whole process. Policies not subject to binding commitments and which may change with time are called time-inconsistent. With a change in government, attitudes towards liberalization and monetary policy may change radically. Economic participants may act on the assumption that a policy change away from liberalization could eventually take place, giving rise to a degree of psychological resistance. Liberalization becomes more difficult owing to the obstacles placed in its way by economic units acting on these assumptions. Clear policy statements from a truly autonomous Central Bank can considerably mitigate the problem of time inconsistency. In Malta, considerable regulatory powers have recently shifted from the Ministry of Finance to the Central Bank, although the degree of Central Bank independence remains a matter of debate.

Difficulties related to market knowledge

Whether the market will operate properly once liberalized is an important issue in this regard. In other words, how can one know whether the prevailing 'liberalized' interest rate is at its equilibrium level? Leite and

Sundararajan (1990) discuss five indicators which can help the policy-maker determine whether interest rates are grossly out of equilibrium. First, real interest rates must be positive. Second, world interest rates, in relation to domestic rates, allowing for exchange rate variations, should not have a destabilizing effect. Third, the rate of return on investment must be higher than the cost of borrowed funds. Fourth, interest rates in informal markets may indicate whether those in the formal markets are too high or too low. Finally, the relative price of capital and labour must not distort the market for these factors of production.

These indicators (possibly with the exception of the first) are very difficult to measure in a meaningful way, and may not be very useful in practice. To further complicate matters, the ideal, freely operating market is very difficult, if not impossible, to achieve, and there is no guarantee that the market by itself will actually establish an equilibrium rate of interest.

Strategic Options

Villanueva and Mirakhor (1990) argue that the process of liberalization, if not properly prepared for, and if not properly phased, could generate adverse consequences including sharp and sudden rises in interest rates, bankruptcies of financial institutions and loss of monetary control. They point to three factors that should be considered in order to attain desirable results from the process of liberalization. These are stable macro-economic conditions; strong and effective bank supervision; and graduality in the pace of removal of controls.

Macro-economic instability may lead to financial instability, which in turn may lead to financial collapse of a large number of firms. Coupled with weak bank supervision, instability may foster unsound banking practices, where high risk is assumed and high interest rates are charged, thus further fuelling instability.

Villanueva and Mirakhor define bank supervision in respect of such policies as having adequate reserves against loan losses, adequate bank capitalization, limits of bank exposures to shareholders and large borrowers, limits on foreign exchange exposure, a deposit insurance scheme with appropriate costs that reflect risk, adequate number of bank examiners and supervisors, and absence of political interference with the enforcement of bank supervisory and regulatory controls. They suggest strategies of interest rate liberalization in line with the degree of macro-economic stability and adequate bank supervision prevailing in the country. They identify four possible scenarios in this regard: stable and unstable macro-economic environment (SM and UM) and adequate and inadequate bank supervision (AS and IS).

The best possible situation is of course the SM/AS combination, where interest rate liberalization may be implemented immediately. At the other extreme, there is the UM/IS combination, which is the worst possible situation. Under the latter circumstances, experience shows that rapid interest rate liberalization leads to excessively high interest rates and economic turmoil. Villanueva and Mirakhor cite the experience of Chile, Argentina and Uruguay to back this assertion.

The UM/AS combination allows for the commencement of gradual interest rate liberalization, while the SM/IS combination calls for the enhancement of bank supervision before liberalization. Villanueva and Mirakhor cite the favourable experiences of Korea, Sri Lanka and Indonesia to explain the end result of gradual liberalization under the last two scenarios.

Options for Malta

The strategy options suggested by Villanueva and Mirakhor are directly applicable to Malta.

It should be stressed at this juncture that the role of the Maltese monetary authorities is of paramount importance in whatever option is chosen. An important requisite is that the authorities must be credible and transparent in their intentions to liberalize, possibly by adhering to a strict pre-announced timetable of deregulatory steps. Such a timetable has actually been issued by the Maltese government, in conjunction with its application for accession to the European Union, and to a large extent it is being complied with.

Monetary policy should be clearly separated from fiscal expediency, a condition which calls for an independent Central Bank, and changes have already been made towards this end, since the government is at present committed not to resort to Central Bank credit. Moreover, effective instruments of monetary control need to be introduced, and this is being carried out in Malta by means of the repo auctions, described above, through which the Central Bank influences short-term interest rates.

Another requisite in the Maltese context is that the government's budget deficit and its dominance of domestic credit need to be reduced drastically.

There must also be a strong commitment by the banks to attain efficiency and as much as possible refrain from acting oligopolistically in preparation for the ushering in of a liberalized structure.

Another important requirement in this regard is that bank supervision and prudential regulation continue to be strengthened and brought in line with EU directives.

The next challenge for financial deregulation will be the complete liberalization of exchange control and the establishment of a market for the Maltese lira, in line with the requirements for European Union membership. In turn this will require balance of payments stability, calling for tighter fiscal and monetary policies.

Conclusion

This chapter has looked at the advantages and disadvantages of interest rate liberalization. Its main argument is that in theory reliance on the market mechanism would result in an interest rate that tended to move continually towards equilibrium, thereby fostering better allocation of resources and efficiency.

It has been pointed out, however, that markets are never perfect, and imperfections may create distortions which work against efficient allocation of resources. This argument need not be against liberalization; on the contrary, it calls for the identification and minimization of such imperfections.

The implications of this line of argument with regard to Malta are that changes towards a fully liberalized money market in Malta should not be undertaken without due consideration of market imperfections and the economic and supervisory realities prevailing in the country. For this reason, the present authors agree with the policy which has been adopted by the Maltese monetary authorities: to liberalize interest rates gradually and in regulated and supervised stages.

Steps have already been taken to transfer key monetary controls from government to a more autonomous Central Bank. The Central Bank has in turn replaced direct controls with market-oriented instruments of interest rate management with the aim of eventually attaining complete interest rate deregulation. The liberalization of external capital flows and reforms leading to the establishment of a market for the Maltese lira remain the main challenge to be faced in the coming years.

Acknowledgements

The authors would like to acknowledge the useful suggestions put forward by Gordon Cordina in the preparation of this chapter.

References

Azzopardi, P.V. (1991). 'Interest Rate Liberalisation in Malta.' *Bank of Valletta Review*, no. 4, autumn, pp. 1–16.

Briguglio, L. (1989) 'Currency Demand and the Underground Economy in Malta.' Paper presented during an International Conference on Middle East Business and Economic Research, University of Cairo, January.

Central Bank of Malta (1984) 'Interest Rate Developments in Malta 1980–1984.' *Central Bank Quarterly Review*, Malta: September, pp. 20–33.

Central Bank of Malta (1991) 'Commercial Bank Loans and Advances Outstanding by Main Sector.' *Central Bank Quarterly Review*, Malta: March.

Central Bank of Malta (1992–95) *Annual Report*(s). Malta.

Consiglio, J.C. (1991) 'Increasing the Rate and Improving the Scope of Personal Savings in Malta.' *Bank of Valletta Review*, no. 3, spring, pp. 17–26.

Cordina, G. (1992) 'Monetary Control Perspectives: Aims and Prospects for the Maltese Economy.' BA (Hons Economics) dissertation, University of Malta.

Cutajar, T. (1992) 'Interest Rate Liberalisation in Malta.' BA (Hons, Economics) dissertation, University of Malta.

Dean, J.W. (1993) 'Will Financial Deregulation Be Good for Malta?' *Bank of Valletta Review*, no. 8, autumn, pp. 1–12.

Galdes, A.P. (1989) 'Towards a More Efficient Financial Market.' *Central Bank Quarterly Review*, Malta: December, pp. 35–37.

Galdes, A.P. (1990a) 'Malta's Financial Market in Transition.' *Central Bank Quarterly Review*, Malta: June, pp. 39–41.

Galdes, A.P. (1990b) 'Conditions for Further Development of the Financial Sector in Malta.' *Central Bank Quarterly Review*, Malta: December, pp. 36–40.

Guillaumier, M.A. (1991) 'The Financial Sector and Economic Development in Malta.' *Central Bank Quarterly Review*, Malta: March, pp. 31–39.

Khatkhate and Villanueva, D. (1978) 'Operations of Selective Credit Policies in Developing Countries.' *World Development*, vol. 6, no. 7/8: pp. 979–990.

Leite, S.P. and Sundararajan, V. (1990) 'Issues in Interest Rate Management and Liberalisation.' *IMF Staff Papers*, vol. 37, no. 4, Washington, DC: IMF.

McKinnon, R.I. (1973) *Money and Capital in Economic Development*. Washington: Brookings Institution.

Malta, Government of (1987) Legal Notice 81 of 1987: 'Interest Rates on Loans and Advances Order, 1987.'

Malta, Government of (1991–5) Budget speeches by the Minister of Finance. Malta: Ministry of Finance.

Mas, I. and Talley, S.H. (1990) 'Deposit Insurance in Developing Countries.' *Finance and Development*, vol. 27, no. 4, pp. 43–45.

Shaw, E.S. (1973) *Financial Deepening in Economic Development*. London: Oxford University Press.

Tabone, J. (1991). Speech on the Inauguration of a BOV bank branch as reported in the *Sunday Times* (Malta) by Staff Reporter, 'BOV Chairman Urges Interest Rate Liberalisation,' 26 May.

Tarling, R.J. and Rhodes, J. (1989) *A Study of the Economic Impact of Tourism on the Economy of Malta*. Malta: P.A. Cambridge Consultants for the Ministry of Tourism.

Vassallo, F.J. (1994a) 'A Framework for Monetary Policy in Malta.' *Central Bank Quarterly Review*, Malta: June, pp. 47–53.

Vassallo, F.J. (1994b) 'Address to the Chartered Institute of Bankers.' *Central Bank Quarterly Review*, Malta: December, pp. 49–53.

Vassallo, F.J. (1995) 'The Challenges Ahead.' *Central Bank Quarterly Review*, Malta: December, pp. 65–69.

Villanueva, D. and Mirakhor, A. (1990) 'Strategies for Financial Reforms: Interest Rate Policies, Stabilisation, and Bank Supervision in Developing Countries.' *IMF Staff Papers*, vol. 37, no. 3. Washington, DC: IMF, pp. 509–536.

12

The Money Supply Process in Two Small Island States: Malta and Cyprus, 1960–1993*

Joe Falzon

> What, then, has determined and will determine the value of the franc? First, the quantity, present and prospective, of the francs in circulation. Second, the amount of purchasing power which it suits the public to hold in that shape. . . . The first of these two elements . . . depends mainly on the loan and budgetary policies of the French Treasury. The second of them depends mainly . . . on the trust or distrust which the public feel in the prospects of the value of the franc.
>
> John Maynard Keynes, *A Tract on Monetary Reform* (1923)

The money supply process is of considerable importance in the economy because of the close relationship between changes in the money supply and changes in real output. Friedman and Schwartz (1963), in their celebrated study of the monetary history of the United States, showed that there exists a positive correlation between the money supply and real output: increases in the total stock of money occurred during expansions, while decreases were witnessed during recessions.

Moreover, Sims (1972) showed that the innovations in the money stock preceded the innovations in output. Past changes in real output were unable to predict changes in the money stock. On the other hand, past changes in the money stock could help predict changes in output and in real economic activity. More recently, Stock and Watson (1989), using detrended data concerning the money stock, also present support for the view that movements in the quantity of money help to forecast changes in real output.

---

* An earlier version of this chapter was presented as a paper at the 1994 Annual Conference of the European Association of University Teachers of Banking and Finance, held in Modena, Italy, 8–10 September 1994.

Becketti and Morris (1992) have investigated the ability of money to predict economic activity after the 1970s. They found that, except during the early 1980s, money has remained a useful indicator of future economic activity. A number of developments in financial markets during the early 1980s changed the relationship between money and future real growth, thereby reducing the ability of money to forecast economic activity. When this period of change is excluded from the analysis, however, the predictive ability of money to forecast real economic activity remains intact.

As regards monetary policy, Morgan (1993) finds evidence of asymmetric effects: tight monetary policy slows the economy more than easy monetary policy accelerates it. When monetary policy was tight in 1988 and 1989, the US economy seemed to slow in response. Yet when monetary policy was eased in 1990, the economy did not respond accordingly.

Duca (1993) shows how an increased regulatory burden on banks has encouraged households and firms to bypass the banking system in favour of non-bank financial liabilities and assets. These portfolio shifts by investors lead to cases of 'missing money' depicted by unusual weakness in one of the monetary aggregates, and to declines in the role of banks in providing credit.

The importance of the banking system is further underlined when one considers that Cagan (1965), Sims (1972) and King and Plosser (1984) all find that innovations in output are more strongly linked to innovations in 'inside money' (that part of the money stock consisting of bank deposits) than to innovations in 'outside money' (the money issued by the Central Bank in the form of currency and banks' reserves). The reason for this is that, since inside money represents deposits invested through banks into capital projects, there is a direct link between inside money and real economic activity. On the other hand, outside money, or fiat money created by the Central Bank, represents pieces of paper with no direct link to real production. Consequently, these two forms of money, inside and outside money, end up having very different links to real output and economic activity.

In the two small island states of Malta and Cyprus (see Figure 12.1), most of the above issues relating to the relationship between money and output still need to be researched and analysed thoroughly. This chapter makes an attempt to understand the process and growth of the money supply during the past three decades. It tries to analyse three interrelated issues: (a) In what way has the money supply grown? (b) Why has it grown in the way it did? (c) What was the role that the main factors played in this process?

The chapter develops a model of the money supply multiplier, and hence provides a unified framework with which one can coherently analyse the money supply process in these two economies. First it develops the building blocks of the money supply process by considering some basic ratios of bank deposits to currency in circulation. Then it discusses the possible reasons why the public might have changed its behaviour regarding the proportion

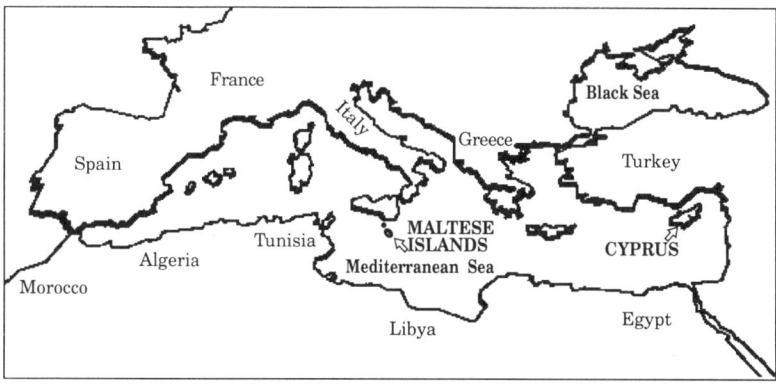

Figure 12.1 The location of Malta and Cyprus in the Mediterranean Sea

of bank deposits and currency it desired to hold. After defining the money supply and the monetary base, the chapter proceeds to develop the algebraic framework for the money multiplier. The ratios developed in the first part of the chapter are combined in order to analyse the two main components of the money multiplier, which is then carefully explained. Next an alternative view of the money supply process is examined by considering the change in domestic credit and net foreign assets, and the effects of an open economy are then incorporated. A final section lists the main results and conclusions of the chapter.

Some Basic Ratios

It is a well-known fact that in Malta there exists quite a large amount of currency in the hands of the Maltese public. Figure 12.2 shows that in Malta, the amount of currency in circulation per capita increased from £67 in 1960 to £1,649 at the end of 1993, while in Cyprus this figure increased from £14 in 1960 to £411 in 1993. By the end of 1993, the currency in circulation per capita in Malta was four times as large as that in Cyprus. It is a less-known fact that the ratio of time deposits to currency in circulation and the ratio of savings deposits to currency have changed considerably during the past 34 years.

As Figure 12.3 shows, the time deposit/currency ratio in Malta decreased continuously from 1.28 in 1971 to 0.64 in 1981. This ratio implies that, whereas in 1971 there was Lm 1.28 in time deposits for every Lm 1 in currency, by 1981 only Lm 0.64 in time deposits existed for every Lm 1 in currency. However, after that year, the public changed its time deposits/ currency portfolio composition. Whereas from 1971 to 1981 the public held continuously smaller and smaller proportions of time deposits to currency in their portfolios, in 1981 this behaviour suddenly changed. The ratio of time

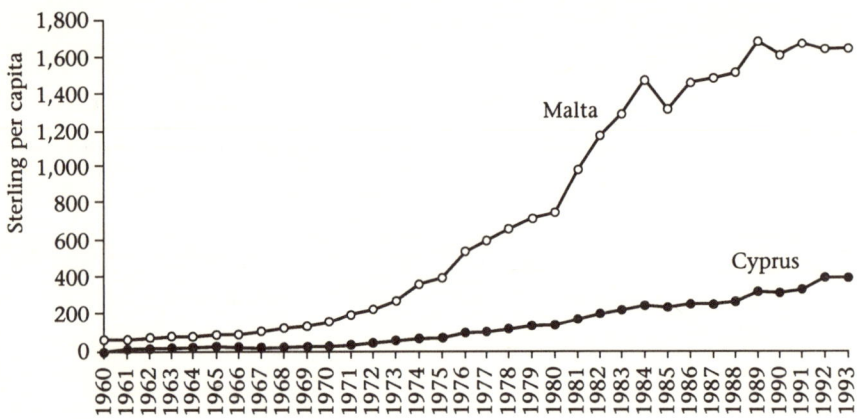

Figure 12.2   Currency per capita in sterling

deposits to currency held by the public doubled from its lowest level of 0.64 in 1981 to 1.30 by September 1992.

A similar change in behaviour occurred with respect to savings/currency holdings. Figure 12.3 also depicts how the savings deposits/currency ratio declined from 0.85 in 1969 to its lowest level of 0.36 in 1983. Then it shot up to its record level of 1.008 in 1992. Hence, in 1983–84, another drastic change in the public's preferences occurred in the desired savings/currency portfolio composition.

As Figure 12.3 also shows, the path of the demand deposit/currency ratio followed two cycles. In the first phase, it decreased from 0.36 in 1969 to 0.10

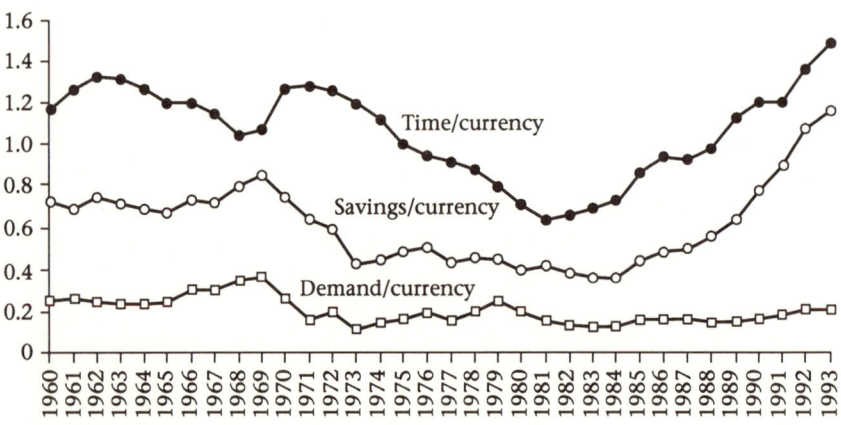

Figure 12.3   Deposits to currency ratios in Malta
Source: Data obtained from various issues of the Central Bank of Malta *Quarterly Review*, the Central Bank of Cyprus *Quarterly Review* and *International Financial Statistics*, Malta and Cyprus pages (IMF, various issues)

in 1973, and increased again to 0.24 in 1979. Then, in the second cycle, it once again dropped to 0.125 in 1983, to then rose to 0.21 by 1992.

A common trend emerges in all these three ratios. The ratios of bank deposits to currency in Malta first declined to very low levels and then changed direction, to increase continuously thereafter. Figure 12.5 (below) shows the proportion of currency in broad money in both Malta and Cyprus.

Figure 12.4 depicts the ratios of bank deposits to currency in Cyprus. A significant change occurred in the time deposit/currency ratio in 1983. This ratio fluctuated slightly around 4.02 between 1973 and 1983 but increased continuously each year to reach the large figure of 10.01 by 1993. The public in Cyprus also changed its portfolio composition of domestic monetary assets after 1983. Each year the public increased its holdings of time deposits, from four times as much as currency in 1983 to ten times as much as currency holdings by 1993.

The changes in the other two ratios in Cyprus were not so drastic as the change in the time/currency ratio. There was a slight decline in the demand/currency ratio between 1.31 in 1970 and 0.76 in 1975, and thereafter a small increase each year that reached 1.39 by 1993. The savings/currency ratio fluctuated very closely around an average of 0.74 between 1976 and 1991, and thereafter declined to 0.66 by 1993. In Cyprus, the main phenomenon that has occurred is the large and significant increase of 150 per cent in the time deposit/currency ratio between 1983 and 1993. Moreover, the high level of this ratio attained after 1988 (reaching 10.01 by 1993) completely dwarfs the levels (1.39 and 0.66) of the other demand and savings/currency ratios.

The proportion of currency in broad money (the sum of currency, demand, savings and time) in both Malta and Cyprus is shown in Figure 12.5. Throughout the whole period, the public in Malta held a higher proportion of

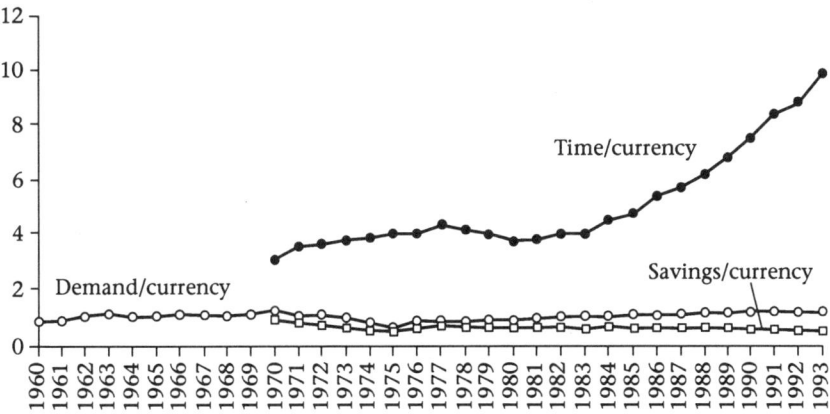

Figure 12.4 Deposits to currency ratios in Cyprus

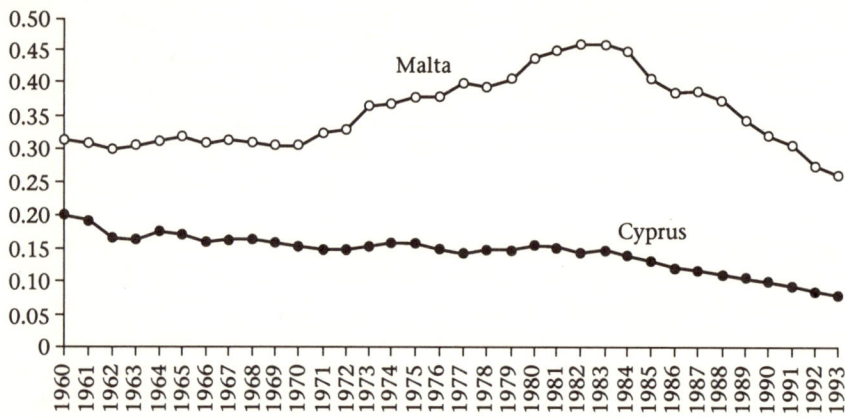

Figure 12.5   Ratio of currency to broad money

currency in its portfolio of domestic monetary assets than the public in Cyprus. From 1960 to 1970, the proportion of currency in the public's portfolio of broad money in Malta fluctuated around 31 per cent, while in Cyprus it declined from 20 to 15.3 per cent. Between 1970 and 1983, the proportion of currency in broad money in Cyprus remained almost constant at an average figure of 14.9 per cent, while in Malta it increased continuously each year from 30.5 per cent in 1970 to 46 per cent in 1982. This means that in 1982, currency in Malta amounted to 46 per cent of the total value of all domestic monetary assets.

After 1983, the proportion of currency in broad money declined constantly in both countries. In Cyprus, the proportion of currency in broad money decreased by half from 14.6 per cent in 1983 to 7.6 per cent in 1993, while in Malta the proportion of currency also fell by almost half from 46 per cent in 1982 to 25.8 per cent in 1993. The values for 1993 in both countries represent all-time low levels for the proportion of currency in the respective local monetary assets.

Explaining the Change in the Public's Behaviour

There are several factors that might induce the public to change its behaviour as regards its bank deposits/currency holdings. Mishkin (1992) lists two main factors that influence the demand for an asset such as currency or bank deposits: changes in people's wealth and changes in the expected return on one asset relative to the expected return on alternative assets.

The wealth elasticity of time and savings deposits is higher than that of currency. Consequently, as wealth grows, the demand for time and savings deposits will grow at a faster rate than the demand for currency. Hence, one would expect to find a positive correlation between wealth and the time deposit/currency and savings deposit/currency ratios.

The Maltese economy registered significant growth rates in real output and income during the 1970s. Hence, the wealth elasticity effect should show significant growth in the demand for time and savings deposits *vis-à-vis* currency demand. This implies rising time/currency and savings/currency ratios, contrary to what is depicted in Figure 12.3. Hence, during the 1970s, the wealth effect either was not significant enough or was offset by other, stronger forces affecting these ratios in the opposite way.

Expected returns on bank deposits relative to currency can be influenced by such factors as the interest rate received on bank deposits (the opportunity cost of holding currency), interest rates on alternative assets, bank panics and uncertainty, and perceived high income taxes that fuel the underground economy.

As the interest rate on bank deposits increases, people will desire to hold less currency (which pays zero interest) and more bank deposits. Hence, the time and savings deposit/currency ratios should be positively related to the interest rate paid on bank deposits. In the United States, until 1980, banks were restricted by Regulation Q to a maximum interest rate on time deposits that did not allow their interest rate to fluctuate. Similarly, interest rates in Malta were fixed at 3 per cent for savings and 5 per cent for time deposits throughout the 1970s and 1980s except for fixed deposits, which enjoyed 6 per cent during 1982–84 and 5.5 per cent in 1985–86. Hence, one cannot use changes in interest rates as an explanation for the variable time and savings/currency ratios that we witness in Figure 12.3.

The demand for time deposits is also sensitive to changes in interest rates on alternative assets such as government bonds or foreign-currency deposits. When interest rates on alternative assets rise above the maximum 5 per cent allowed on time deposits, the expected return on time deposits would fall in relation to these other assets. The demand for time deposits would slow down relative to currency, decreasing the time deposit/currency ratio. During the late 1980s, the expected return both on foreign-currency deposits (given the high foreign interest rates and fixed domestic exchange rate) and on domestic property (given the local boom in real estate) was definitely larger than 5 per cent per annum. Hence one would expect the demand for fixed and savings deposits to slow down because the public would prefer to hold these alternative assets. This would cause the time deposit/currency ratio and savings deposit/currency ratio to fall. Contrary to our expectations, however, as depicted in Figure 12.3, both these ratios increased considerably between 1987 and 1992.

Panics and uncertainty in the banking system can have a devastating effect on the expected returns of holding bank deposits. People lose confidence in banks and no longer perceive them to be a safe place in which to deposit their hard-earned, and maybe lifelong, savings. Depositors will withdraw their bank deposits *en masse*, causing a run on the banks, deeper bank panic and uncertainty, and a sharp decline in time and savings/currency ratio. In

Malta, however, apart from a certain increase in uncertainty due to the changing ownership structure of National Bank of Malta in 1973 and Barclays in 1976, no outbreaks of panics and runs on the banking system at large were witnessed in the time period analysed. Hence one cannot use this factor as an explanation for the changing time and savings/currency ratios.

Finally, when people perceive income tax rates to be excessively high (up to 1989, the highest income tax rate was 65 per cent), they have a considerable incentive to conduct transactions and receive payment only in cash. As Figure 12.2 shows, there was almost Lm 1,000 per person in currency in circulation in 1992 – significant evidence of the large underground economy in Malta. Higher taxes will lead to an increase in currency relative to bank deposits and hence to a drop in time/currency and savings/currency ratio as witnessed in Figure 12.3 during the 1970s. However these ratios *increased* during the 1980s, implying either lower income taxes or greater honesty in income tax reporting. The income tax structure was changed in 1990 by reducing the highest income tax rate from 65 to 35 per cent and by allowing married couples to file separate income tax returns. Moreover, there is no proof that the size and extent of the underground economy decreased during the 1980s. Hence, one cannot use tax avoidance as an explanation for the changing deposit/currency ratios as depicted in Figure 12.3.

It seems that the potential factors listed above – wealth elasticity, changes in interest rates on domestic bank deposits, changes in expected returns on alternative assets, bank panics and tax avoidance – are not able to explain consistently the changes that occurred in Malta in time and savings/currency ratios during *both* the 1970s *and* the 1980s. It seems that there is still need for further research to develop a consistent theory to help explain the changes in these deposit/currency ratios.

Definition of Money Supply and Monetary Base

The different ratios of bank deposits to currency, analysed in the above two sections, form the basis of the money supply multiplier as derived in this chapter. The public affects the money multiplier by its decision as to how much it decides to hold in bank deposits relative to currency. The larger the ratio of bank deposits to currency that the public holds, the larger the money multiplier and the total money supply.

The definition of money supply that we use is M2 or broad money, which is the sum of currency in circulation (C), demand (checkable) deposits (D), savings deposits (S) and time deposits (T):

$$M2 = C + D + S + T \tag{12.1}$$

The objective of this chapter is to explore the nature of the money multiplier: the relationship between M2 and the monetary base (MB). That is, we want to know the nature of m^* in:

$$M2 = m^* . MB \tag{12.2}$$

Early multiplier studies were developed by Brunner (1961) and Brunner and Meltzer (1964). Andersen and Jordan (1968) and Burger *et al.* (1971) developed multiplier relationships between the monetary base and M1 (currency plus checkable deposits). An extensive framework for the determination of the money supply was developed by Boorman and Havrilesky (1972), while Johannes and Rasche (1979) built a detailed multiplier model that incorporated both demand and time deposits. More recent treatments include those of Garfinkel and Thornton (1991) and Mishkin (1992).

The monetary base consists of currency in circulation plus banks' reserves (vault cash plus deposits with the Central Bank):

$$MB = C + BR \tag{12.3}$$

Several authors, including Friedman (1959) and Lothian (1976), refer to the monetary base as 'high-powered money'. Others, like Gurley and Shaw (1960), use the term 'outside money' – an obligation of the government or the Central Bank that is outside the control of the private sector, in contrast to 'inside money', which is an obligation of the private sector.

Meulendyke (1990) describes extensively the possible roles for the monetary base as operating and intermediate targets of monetary policy. The final objective of monetary policy is to reach desirable levels in its 'ultimate targets', usually the price level and real output. As immediate 'instruments' of monetary policy, the Central Bank makes use of open-market operations, the discount rate and required reserve ratios. Through these instruments, the Central Bank succeeds in controlling very closely its 'operating targets' such as non-borrowed reserves, non-borrowed monetary base and short-term money market interest rates. Movements in these operating targets significantly influence changes in the 'intermediate' targets of monetary policy. These intermediate targets, such as measures of the money supply and bank credit, are intermediate between the instruments and operating targets which the Central Bank can tightly control, and the ultimate targets that can be influenced only indirectly.

The monetary base is usually referred to as an operating target; it is to be controlled not for its own sake but in order to achieve some desired behaviour in another variable, usually some measure of the money supply. The monetary base, however, has also been proposed as an intermediate target in place of the traditional monetary aggregates such as M1 or M2. This is done in order to try to solve the potential problems associated with controlling these monetary aggregates in the short run or even to treat the monetary base itself as a narrow monetary measure.

The monetary base forms part of the Central Bank liabilities, which by definition always change by the same amount as Central Bank assets. When the Central Bank, for example, buys Lm 1 million worth of foreign currency or Lm 1 million worth of domestic government bonds, the Central Bank's assets will increase by Lm 1 million. This, by definition, will increase the Central Bank liabilities by Lm 1 million too.

In the case of Malta, the monetary base accounts, on average, for 75 per cent of total liabilities of the Central Bank. A Lm 1 million increase in Central Bank assets, while increasing Central Bank liabilities by the same amount (Lm 1 million), will normally increase the monetary base by Lm 0.75 million. The rest goes into 'other (residual) liabilities'.

Hence, total liabilities (L) equals monetary base (75 per cent) plus residual liabilities (RL) (25 per cent):

$$L = MB + RL \tag{12.4}$$

However, to capture unexpected shifts in Central Bank liabilities from monetary base to residual liabilities and vice versa, we develop the multiplier in terms of total liabilities of the Central Bank:

$$M2 = m^* (MB + RL) = m^* . L \tag{12.5}$$

This will also be helpful because the initial change originates in the Central Bank asset side, which by definition will always equal the change in total liabilities.

We want to find out by how much M2 will increase because of a Lm1 million increase in Central Bank assets (A), which will increase Central Bank liabilities (L) by Lm 1 million. That is, we want to find m^* in

$$M2 = m^* . A = m^* . L \tag{12.6}$$

Derivation of the Money Multiplier

The multiplier presented here is different from the simple money multiplier (1/rr) presented in textbooks such as those of Baumol and Blinder (1982) and Mayer *et al.* (1987). This simple multiplier, which is the inverse of the reserve requirement on checkable deposits, is built on two unrealistic assumptions: that banks hold no excess reserves and that the public holds no extra currency in circulation. When a member of the public deposits a sum of money in a bank, the bank manager is assumed to hold a proportion of this sum as required reserves but lends out all the rest. Hence, excess reserves are assumed to be kept at zero all the time. Moreover, whenever person A who receives the loan spends it, and person B receives the proceeds, person B is assumed to deposit the full amount into his/her bank without withdrawing any extra currency. Hence no leakage into extra currency will occur. These two assumptions are unrealistic because banks do hold excess reserves and because we do witness leakages that increase the currency in circulation.

More realistic models of the money multiplier need to adjust these two assumptions to take into account the facts of increasing currency and changing bank reserves. Thomas (1986) and Mishkin (1992) develop a more realistic model of the money multiplier using the ratios of both the currency to checkable deposits and the time deposits to checkable deposits, as well as the required reserve ratios of checkable and time deposits. Thornton (1992) and Mishkin (1992) also extend the analysis to derive a money multiplier for M2.

The money multiplier developed in this chapter follows the approach of Freeman (1992), who builds a money multiplier around the ratio of bank deposits to nominal fiat money. Private individuals affect the money multiplier by their decision as to how much money to hold in currency and deposits respectively. The more the public holds deposits relative to currency, the larger the money multiplier and the total money supply.

The objective is to find out the relationship between the money supply (M2) and the level of Central Bank liabilities (L), which by definition are always equal to Central Bank assets (A).

$$M2 = m^* . L = m^* . A \tag{12.6}$$

The money multiplier m^* tells us by how much the money supply will increase for a unit increase in Central Bank assets (liabilities).

We know that broad money (M2) is equal to currency in circulation (C) plus checkable deposits (D) plus savings deposits (S) plus time deposits (T). Thus,

$$M2 = C + D + S + T \tag{12.1}$$

Central Bank liabilities are equal to money base (MB) plus residual liabilities (RL):

$$L = MB + RL \tag{12.4}$$

Monetary base (MB) is equal to currency public (C) plus bank reserves (BR). Hence:

$$MB = C + BR \tag{12.3}$$

Substituting (12.3) into (12.4), we get

$$L = C + BR + RL \tag{12.7}$$

Then we substitute (12.1) and (12.7) into (12.6) to get

$$C + D + S + T = m^*(C + BR + RL) \tag{12.8}$$

Dividing throughout by C,

$$1 + (D/C) + (S/C) + (T/C) = m^*(1 + (BR/C) + (RL/C)) \tag{12.9}$$

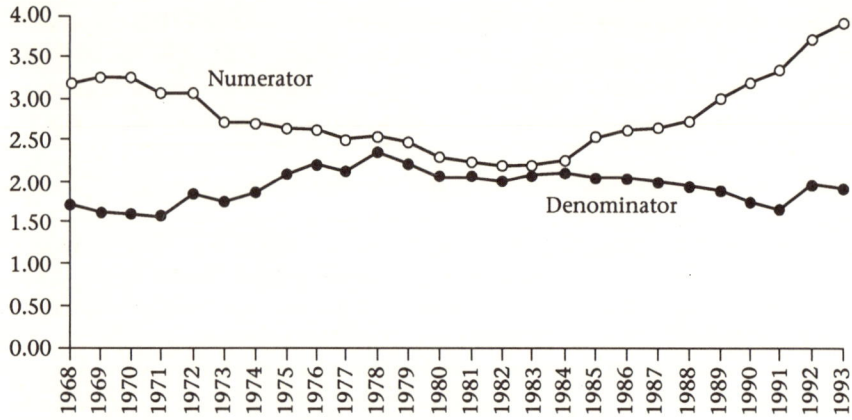

Figure 12.6   Components of multiplier in Malta

Hence we get the money multiplier as:

$$m^* = \frac{1 + (D/C) + (S/C) + (T/C)}{1 + (BR/C) + (RL/C)}$$ (12.10)

where D/C is the ratio of checkable deposits to currency, S/C is the ratio of savings deposits to currency, T/C is the ratio of time deposits to currency, BR/C is the ratio of banks' reserves to currency, and RL/C is the ratio of residual liabilities to currency.

The values of the numerator and denominator of the multipliers in Malta and Cyprus are shown in Figures 12.6 and 12.7 respectively, while the value of the multiplier m^* in both countries is depicted in Figure 12.9.

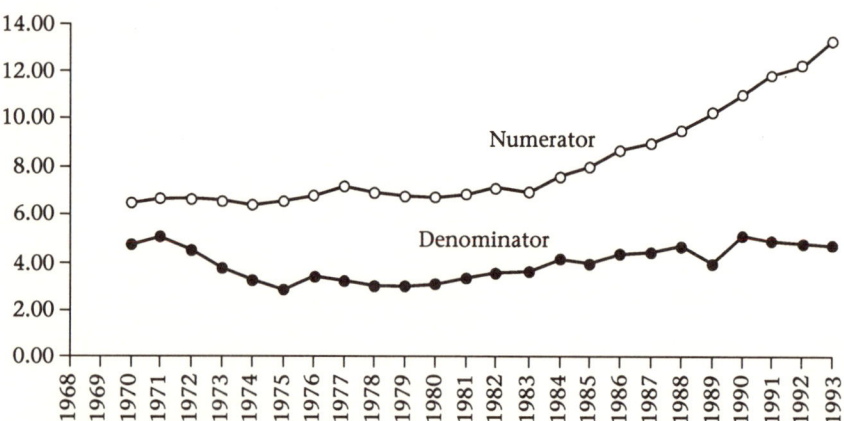

Figure 12.7   Components of multiplier in Cyprus

The Components of the Multiplier

The *numerator* of the money multiplier, 1 + (D/C) + (S/C) + (T/C), is simply the sum of the ratios of time deposit/currency, savings deposit/currency and demand deposit/currency described earlier in this chapter. In Malta, the relative increase of currency as a proportion of broad money between 1969 and 1982, as witnessed in Figure 12.5, is shown in Figure 12.6 as a drop in the numerator of the money multiplier from 3.29 in 1969 to 2.18 in 1982.

In 1982, we witnessed in Malta the start of the relative shift out of currency into fixed, savings and demand deposits. This caused the numerator of the money multiplier to increase continuously from the low level of 2.18 in 1982 to its highest level of 3.87 in 1993.

The *denominator* of the multiplier, 1 + (BR/C) + (RL/C), shows the relative proportion of bank reserves and residual liabilities in the central bank total liabilities. The denominator is the inverse of the proportion of currency held by the public in the central bank total liabilities, as depicted in Figure 12.8. The value of the total denominator of the money multiplier, 1 + (BR/C) + (RL/C), in Malta, as depicted in Figure 12.6, increased from 1.55 in 1971 to the highest level of 2.31 in 1978. Thereafter, it fell again to 1.62 by 1991.

The ratio of currency to Central Bank liabilities in Malta is depicted in Figure 12.8. The proportion of currency in Central Bank liabilities fell from 64 per cent in 1971 to 43 per cent in 1978, and then increased again to 61 per cent by 1991. The increase in the denominator of the multiplier, 1 + (BR/C) + (RL/C), between 1971 and 1978, as shown in Figure 12.6, corresponds to the decline of the currency/Central Bank liabilities (C/L) ratio depicted in Figure 12.8. Consequently, the decline of the denominator between 1978 and 1991 corresponds to the increase of the currency/liabilities ratio shown in Figure 12.8. Hence, in the last phase, while the proportion of banks' reserves and residual liabilities has been going down, the proportion of currency held

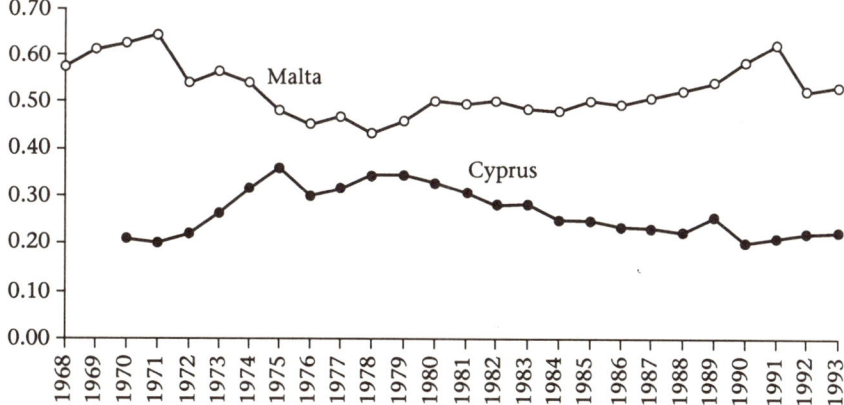

Figure 12.8 Ratio of currency to central bank's liabilities

by the public in Central Bank total liabilities has been progressively increasing.

In Cyprus, the almost constant ratio of currency to broad money of 0.148 between 1970 and 1983, as depicted in Figure 12.5, reflects itself as an almost constant *numerator* of the multiplier in Cyprus of an average value of 6.71, as shown in Figure 12.7. Between 1983 and 1993, the proportion of currency to broad money in Cyprus, as witnessed in Figure 12.5, decreased from 14.57 to 7.64 per cent. This reflected itself in a significant increase in the numerator of the multiplier in Cyprus, as shown in Figure 12.7, from a value of 6.86 in 1983 to a figure almost double in size of 13.07 by 1993.

The *denominator* of the multiplier in Cyprus, also depicted in Figure 12.7, registered a decline from a value of 4.76 in 1970 to a value of 2.79 in 1975, and thereafter increased slowly to regain a value of 4.87 by 1990. This reflects the reciprocal movement of the ratio of currency to Central Bank liabilities in Cyprus, shown in Figure 12.8, which first increased from 0.21 in 1970 to 0.36 by 1975, then declined to 0.205 again by 1990, and afterwards increased slightly to 0.22 by 1993.

There is a simple relationship between the value of the numerator of the money multiplier and the proportion of currency in broad money. We know that

$$M2 = C + D + S + T \tag{12.1}$$

So

$$(M2/C) = 1 + (D/C) + (S/C) + (T/C) \tag{12.11}$$

Hence the value of the numerator of the multiplier, $1 + (D/C) + (S/C) + (T/C)$, as shown in Figure 12.6 for Malta and Figure 12.7 for Cyprus, is simply the reciprocal of the proportion currency in broad money depicted in Figure 12.5.

Moreover, there is also a simple relationship between the denominator of the multiplier and the proportion of currency in total Central Bank liabilities. We know that

$$L = C + BR + RL \tag{12.7}$$

So

$$(L/C) = 1 + (BR/C) + (RL/C) \tag{12.12}$$

Hence the denominator of the money multiplier, $1 + (BR/C) + (RL/C)$, depicted in Figures 12.6 and 12.7, is simply the reciprocal of the proportion of

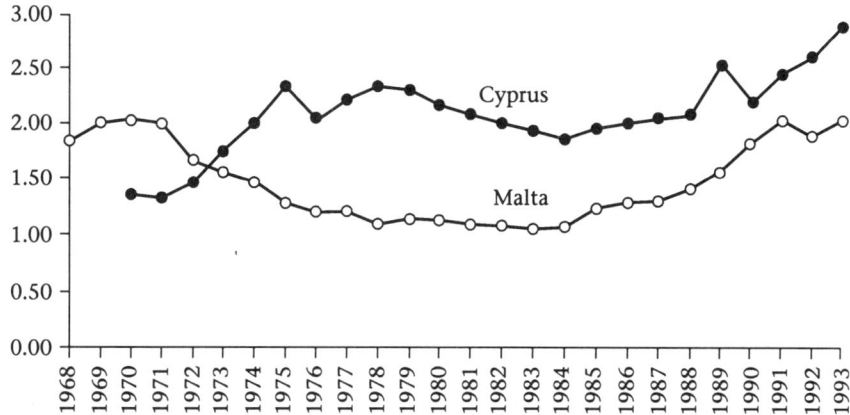

Figure 12.9 Multiplier

currency in Central Bank liabilities as shown in Figure 12.8 for both countries.

The Money Multiplier

The money multiplier is finally derived by dividing the numerator $1 + (D/C) + (S/C) + (T/C)$ by the denominator $1 + (BR/C) + (RL/C)$ as required in equation (12.10).

The path of the money supply multiplier between 1968 and 1993 is shown in Figure 12.9. The money multiplier for Malta followed the path of a valley passing through three phases: a phase of decline, a constant-level phase, and a phase of increase. The money multiplier for Cyprus also followed a path of three phases: a phase of increase, a phase of decline, followed by another phase of increase. The years 1978 and 1984 happen to be the cut-off years between these three phases for both Malta and Cyprus.

The value of the money multiplier m^* changed through the three phases in the manner shown in Tables 12.1–12.3 respectively.

Table 12.1 Changes in the value of the money multiplier for Cyprus and Malta, 1970–78

Year	Cyprus	Malta
1970	1.37	2.05
1971	1.33	1.98
1972	1.48	1.65
1973	1.77	1.55
1974	2.02	1.47
1975	2.33	1.28
1976	2.04	1.19
1977	2.23	1.18
1978	2.35	1.09

Table 12.2   Changes in the value of the money multiplier for Cyprus and Malta, 1978–84

Year	Cyprus	Malta
1978	2.35	1.09
1979	2.31	1.13
1980	2.18	1.13
1981	2.07	1.09
1982	2.00	1.08
1983	1.95	1.05
1984	1.87	1.07

Table 12.3   Changes in the value of the money multiplier for Cyprus and Malta, 1984–93

Year	Cyprus	Malta
1984	1.87	1.07
1985	1.96	1.23
1986	2.00	1.28
1987	2.04	1.31
1988	2.07	1.40
1989	2.54	1.57
1990	2.21	1.84
1991	2.48	2.02
1992	2.62	1.91
1993	2.91	2.06

The tables show that during the past ten years, the value of the money multiplier in Malta doubled from 1.05 in 1983 to 2.06 in 1993. This means that whereas, in 1983, a Lm 1 million increase in total Central Bank assets increased broad money (M2) by Lm 1.05 million, in 1993 a similar Lm 1 million increase in total assets produced a Lm 2.06 million rise in broad money.

In Cyprus, in the past ten years, the value of the multiplier increased by 56 per cent from 1.87 in 1984 to 2.91 in 1993. Moreover, since 1973 the multiplier in Cyprus has always been larger than the multiplier in Malta. The gap between the two multipliers, which was largest in 1978, narrowed slowly to the lowest figure in 1990, but started to widen again thereafter. By 1993, the multiplier in Cyprus was 41 per cent larger than the multiplier in Malta.

Domestic Credit and Net Foreign Assets

A second way of analysing the growth in the money supply is by considering the asset side of the consolidated balance sheets of the Central Bank, the commercial banks and other financial institutions.

Morgan (1992) explains why credit matters, and differentiates between the 'money view' and the 'credit view' of the monetary transmission mechanism.

According to the money view, only money matters. Reduced availability of bank loans does not matter because firms can maintain their spending by borrowing in the commercial paper and bond markets. According to the credit view, however, bank loans do matter. Many small and less established firms may rely solely on banks for loans. Thus bank loans are 'special' because smaller firms may not be able to find alternative sources of credit.

Further support for the credit view comes from Gertler and Gilchrist (1991), who compile cross-section evidence on small and large firms and show why credit does matter. Small firms are squeezed more than large firms by tighter credit conditions and hence are more responsive to changes in monetary policy.

Robinson (1993) shows how during 1985–92, a period of banking difficulties in the United States, prior movements in bank loans are a significant factor in predicting current M2 growth. As regards the Maltese scenario, Falzon (1989) makes a detailed analysis of the portfolio composition of the local banking sector, showing the changing proportions of banks' assets held as short-term funds (cash and balances with the Central Bank), securities (domestic and foreign) and total advances (overdrafts and loans).

Total domestic credit (DC) (extended to the government and private sector) plus total net foreign assets (NFA) (held by the monetary authorities and commercial banks) is by definition equal to broad money plus 'other items' (OI). Hence

$$DC + NFA = M2 + OI \tag{12.13}$$

Table 12.4 shows the average yearly increases in total domestic credit and net foreign assets for four recent five-year periods in Malta. As can be seen, there was a significant 'structural' shift in the generating process of broad money (M2) in Malta. The increase in net foreign assets was the primary engine of growth of M2 up to 1981. The change in net foreign assets contributed 95 per cent of the total increase in M2 during 1972–76. Domestic credit contributed only 5 per cent. During 1977–81, the share of net foreign

Table 12.4 Average annual increase (Lm million) in domestic credit and net foreign assets for Malta during four five-year periods

	1972–76	1977–81	1982–86	1987–91
Domestic credit	1.6 (5%)	6.2 (14%)	27.3 (69%)	90.0 (91%)
Net foreign assets	33.4 (95%)	37.9 (86%)	12.2 (31%)	9.1 (9%)
Total	35.0 (100%)	44.1 (100%)	39.5 (100%)	99.1 (100%)
of which				
Broad money	29.1 (83%)	41.8 (95%)	36.2 (92%)	84.4 (85%)
Other items	5.9 (17%)	2.3 (5%)	3.3 (8%)	14.7 (15%)

assets fell slightly to 86 per cent, while domestic credit contributed to 14 per cent of the change in broad money.

However, there was a significant change in the money-generating process after 1982. Net foreign assets ceased to be the primary engine of growth of M2. During 1982–86, the change in total domestic credit became the primary factor: 69 per cent of the increase in M2 was caused by domestic credit. The share of net foreign assets fell drastically to 31 per cent.

The reliance on domestic credit as the primary generating process was further intensified during the last five years, 1987–91. Total net foreign assets contributed only 9 per cent of the average annual increase of M2, while the share of domestic credit jumped to 91 per cent. However, since the grand total of the change in domestic credit and change in net foreign assets was split in the proportion 85 per cent into M2 and 15 per cent into 'other items', the average annual increase in domestic credit (Lm 90 million) was larger than the average annual increase in M2 (Lm 84.4 million).

In Cyprus, the increase in domestic credit was the main factor that fuelled the growth in the money supply during the four of the five-year periods under review. Domestic credit accounted for 95 and 91 per cent of the increase in M2 during the first two periods (Table 12.5), while net foreign assets contributed 5 and 9 per cent respectively. During 1982–86, the role of domestic credit in the money-generating process decreased to 82 per cent, while that of net foreign assets increased to 18 per cent. However, during the last five years, 1987–91, the reliance on domestic credit increased again to 95.5 per cent, while that of net foreign assets fell back to 4.5 per cent.

In Cyprus, unlike Malta, there was no 'structural' change in the money-generating process of broad money. Domestic credit was the main engine of money growth while net foreign assets played only a minor part throughout all the four periods under review. Moreover, in three of the four periods (1972–76, 1977–81 and 1987–91) the average annual increase in domestic credit (CY£28.4, CY£68.3 and CY£248.1) was larger than the average annual increase in broad money (CY£25.9, CY£65.8 and CY£232.3).

Table 12.5 Average annual increase (CY£ million) in domestic credit and net foreign assets for Cyprus during four five-year periods

	1972–76	*1977–81*	*1982–86*	*1987–91*
Domestic credit	28.4 (95%)	68.3 (91%)	98.0 (82%)	248.1 (96%)
Net foreign assets	1.6 (5%)	6.6 (9%)	21.3 (18%)	11.7 (4%)
Total	30.0 (100%)	74.9 (100%)	119.3 (100%)	259.8 (100%)
of which				
Broad money	25.9 (86%)	65.8 (88%)	99.4 (83%)	232.3 (89%)
Other items	4.1 (14%)	9.1 (12%)	19.9 (17%)	27.5 (11%)

Effect of an Open Economy

A third alternative way of looking at the growth of M2 is by analysing further the components of the increase in total domestic credit and the flows that result in the final net foreign assets. The change in total domestic credit can be looked at as the sum of the change in bank lending to the private sector and the government funding requirement (part of which is financed through foreign sources). The final change in net foreign assets can be looked at as the increase that occurs through inflows of foreign exchange on the capital account plus the decrease that occurs through outflows on the current account.

Table 12.6 gives the average levels of these quantities in Malta during the past 15 quarters: January 1989 to September 1992. The private sector in Malta borrowed, on average, Lm 20.4 million per quarter; the government had a average deficit of Lm 12.2 million per quarter; capital inflows amounted to Lm 6.8 million; while 'other items' were on average Lm 8.3 million per quarter. Hence, every quarter, on average, Lm 47.7 million worth of monetary assets were initially created.

Of these, however, Lm 23.5 million flowed out through the current account because of the openness of the economy. The final change in M2 that remains, an average of Lm 24.2 million per quarter, is only half the initial increase that would have occurred had no 'money' 'flowed out' of the Maltese economy.

Conclusion

This chapter examined the internal forces of the money multiplier to try to identify the factors that are causing it to change. The chapter first looked at several ratios describing the proportions of bank deposits to currency and the proportions of the Central Bank liabilities to currency. Then it developed a simple algebraic model linking all these ratios, to help determine the money supply multiplier. The last part of the chapter analysed an alternative approach to money supply determination: the domestic credit–net foreign

Table 12.6 Average level per quarter (Lm million), January 1989–September 1992, of various economic quantities affecting Malta's money supply

Change in bank lending to the private sector	20.39	(43%)
Government funding requirement	12.17	(26%)
Capital account balance	6.81	(14%)
Other items	8.34	(17%)
Initial total	47.71	
Current account balance	-23.47	(49%)
Change in broad money (M2)	24.24	(51%)

asset approach. It also incorporated the effects of the openness of the economy.

The first main finding was that people in Malta changed their behaviour during 1981–83 as regards the composition of the monetary assets they would like to hold. They held relatively less currency and more savings and time deposits. This caused the money multiplier to double from 1983 to 1993. In Cyprus, the public also changed its behaviour after 1983 regarding the portfolio composition of its monetary assets. After 1983, the public held relatively much more time deposits and relatively less currency. This caused the money multiplier in Cyprus to increase by 56 per cent from 1983 to 1993.

The second important point was that, in Malta, a structural change in the composition of M2 also occurred during 1982–86. Before 1982, the change in M2 in Malta was caused primarily by changes in net foreign assets. Between 1982–1986, increases in M2 occurred mainly as a result of changes in domestic credit. After 1987, growth of M2 was almost exclusively caused by domestic credit. In Cyprus, however, no such structural change occurred in the money-generating process. The change in domestic credit was the main overriding engine of growth that fuelled the increase in broad money throughout our four five-year periods.

The third main result was that the growth of M2 in Malta that we witness is only half of the initial increase in domestic monetary assets. The other half 'flows out' because of the openness of the Maltese economy. This means that every Lm 2 million of private borrowing from the banking sector will ultimately result in a Lm 1 million increase in broad money (M2).

The value of the multiplier and the underlying money-generating process are important tools of economic policy because they help explain what factors are causing the change and growth in monetary assets. Hence, they can be used to analyse and predict the amount of 'money' available in commercial banks that could be lent to the different sectors of the economy.

Glossary of Terms Used

Currency public = currency in circulation

Currency issued = currency public + banks' cash in vault

Banks' reserves = banks' vault cash + banks' deposits with Central Bank (CB)

Money base = currency issued + banks' deposits with CB

Money base = currency public + banks' cash + banks' deposits with CB

Money base = currency public + banks' reserves

Total CB liabilities = currency issued + banks' deposits CB + other (residual) liabilities

Total CB liabilities = money base + other (residual) liabilities

References

Andersen, L.C. and Jordan, J.L. (1968), 'The Monetary Base: Explanation and Analytical Use.' Federal Reserve Bank of St Louis *Review*, August: pp. 7–11.

Baumol, W.J. and Blinder, A.S. (1982) *Economics: Principles and Policy*, 2nd edition. New York: Harcourt Brace Jovanovich.

Becketti, S. and Morris, C.S. (1992) 'Does Money Still Forecast Economic Activity?' Federal Reserve Bank of Kansas City *Economic Review*, fourth quarter: pp. 65–78.

Boorman, J.T.A. and Havrilesky, T.M. (1972) 'A Framework for the Determination of the Money Supply.' In Boorman and Havrilesky (eds) *Money Supply, Money Demand and Macroeconomic Models*. Arlington Heights, IL: AHM Publishing Corporation, pp. 3–46.

Brunner, K. (1961) 'A Schema for the Supply Theory of Money.' *International Economic Review*, January, vol. 2, no. 1: pp. 79–109.

Brunner, K. and Meltzer, A.H. (1964) 'Some Further Investigations of Demand and Supply Functions for Money.' *Journal of Finance*, May, vol. 19, no. 2: pp. 240–83.

Burger, A.E., Kalish, L. III and Babb, C.T. (1971) 'Money Stock Control and Its Implications for Monetary Policy.' Federal Reserve Bank of St Louis *Review*, October: pp. 6–22.

Cagan, P. (1965) *Determinants and Effects of Changes in the U.S. Money Stock, 1875–1965*. New York: National Bureau of Economic Research.

Central Bank of Cyprus, *Quarterly Review*, various issues.

Central Bank of Malta, *Quarterly Review*, various issues.

Duca, J.V. (1993) 'Regulation, Bank Competitiveness, and Episodes of Missing Money.' Federal Reserve Bank of Dallas *Economic Review*, second quarter: pp. 1–23.

Falzon, J. (1989) 'The Banking Sector: Perspectives and Opportunities.' Department of Economics, University of Malta, August.

Freeman, S. (1992) 'Money and Output: Correlation or Causality?' Federal Reserve Bank of Dallas *Economic Review*, third quarter: pp. 1–8.

Friedman, M. (1959) *A Programme for Monetary Stability*, Millar Lectures, no. 3. New York: Fordham University Press.

Friedman, M. and Schwartz A.J. (1963) *A Monetary History of the United States, 1867–1960*. Princeton, NJ: Princeton University Press.

Garfinkel, M.R. and Thornton, D.L. (1991) 'The Multiplier Approach to the Money Supply Process: a Precautionary Note'. Federal Reserve Bank of St Louis *Review*, July/August: pp. 47–62.

Gertler, M. and Gilchrist, S. (1991) 'Monetary Policy, Business Cycles, and the Behaviour of Small Manufacturing Firms.' National Bureau of Economic Research, Working Paper 3892.

Gurley, J.G. and Shaw, E.S. (1960) *Money in a Theory of Finance*. Washington, DC: Brookings Institution.

International Monetary Fund. *International Financial Statistics*, various issues.

Johannes, J.M. and Rasche, R.H. (1979) 'Predicting the Money Multiplier.' *Journal of Monetary Economics*, vol. 5: pp. 301–325.

King. R. and Plosser, C. (1984) 'Money, Credit and Prices in a Real Business Cycle.' *American Economic Review*, vol. 74: pp. 363–380.

Lothian, J.R. (1976) 'The Demand for High-Powered Money.' *American Economic Review*, vol. 66, no. 1: pp. 56–68.

Mayer, T., Duesenberry, J.S. and Aliber, R.Z. (1987) *Money, Banking and the Economy*, 3rd edition. New York: W.W. Norton.

Meulendyke, A-M. (1990) 'Possible Roles for the Monetary Base.' *Intermediate Targets and Indicators for Monetary Policy*. New York: Federal Reserve Bank of New York, pp. 20–66.

Mishkin, F.S. (1992) *The Economics of Money, Banking and Financial Markets*. New York: HarperCollins.

Morgan, D.P. (1992) 'Are Bank Loans a Force in Monetary Policy?' Federal Reserve Bank of Kansas City *Economic Review*, second quarter: pp. 31–42.

Morgan, D.P. (1993) 'Asymmetric Effects of Monetary Policy.' Federal Reserve Bank of Kansas City *Economic Review*, second quarter: pp. 21–33.

Robinson, K.J. (1993) 'The Relationship between Bank Lending and Money Growth: Were Things Different in the 1980s?' Federal Reserve Bank of Dallas *Financial Industry Studies*, December: pp. 13–26.

Sims, C. (1972) 'Money, Income and Causality.' *American Economic Review*, vol. 62: pp. 540–552.

Stock, J. and Watson, M. (1989) 'Interpreting the Evidence on Money–Income Causality.' *Journal of Econometrics*, vol. 40: pp. 161–181.

Thomas, L.B. Jr (1986) *Money, Banking and Economic Activity*, 3rd edition. Englewood Cliffs, NJ: Prentice-Hall.

Thornton, D.L. (1992) 'Targeting M2: the Issue of Monetary Control.' Federal Reserve Bank of St Louis *Review*, July/August: pp. 23–35.

13

International Banking and Securities Market Regulation: an Analysis of Recent Approaches*

Michael Bowe and Maximilian J.B. Hall

This chapter's main objective is to compare two alternative approaches which have recently emerged for controlling and regulating activities that contribute to market risk in the market for financial services; namely, the European Commission's 'Capital Adequacy Directive' (CAD) (EC, 1993a) and the revised 1995 Basle Committee on Banking Supervision (BC) proposals. These approaches encapsulate two very different strategies to financial market regulation.

The CAD utilizes a standardized measurement framework to prescribe minimum initial capital requirements to cover the market risks faced by both credit institutions[1](henceforth banks) and investment (securities) firms.[2] This followed agreement on earlier proposals attempting to establish a common approach to defining capital and assessing capital adequacy for banks, and the establishment of a minimum capital standard to cover the credit risks faced by banks.[3] The Commission's final proposals for handling the market risks faced by banks and investment firms, in the shape of the CAD (EC, 1993a),[4] were eventually adopted by member states of the European Union in March 1993, for implementation by 1 January 1996.

Regulatory directives such as the CAD impose standards that are public in character and modify the behaviour of financial institutions in an immediate way, through requirements that are imposed prior to, or at least independ-

* A formal version of some of the main arguments of this chapter can be found in 'Rules with or without Discretion: a Comparison of Capital Standards and Market Surveillance as Mechanisms for Regulating Market Risk' (1996) (Economic Research Paper 96–07, Department of Economics, Loughborough University), by the same authors.

ently of, the occurrence of systemic damage. Proponents of such an approach to regulation argue that the standardized nature of such directives enhances the transparency and consistency of regulatory requirements across financial institutions. Its detractors argue that any standardized regulatory framework necessarily incorporates a measurement system which is too simplified adequately to capture the complex interaction of those market risk parameters which contribute to systemic financial crisis. Consistent with this line of thinking is the view that proprietary in-house risk management models may be better able to provide accurate measures of market risk. Therefore, the emphasis in financial market regulation should be directed towards developing more efficient proprietary, in-house practices for managing market risk, supplemented with third-party or market-based supervision and capital standards where required.

One recent influential proposal which incorporates this regulatory approach is that of the Basle Committee (1995). The proposed framework is based upon the establishment of a series of quantitative and qualitative criteria that financial institutions would have to satisfy in order to be able to use their own proprietary systems for the measurement of market risk as the basis for determining a capital requirement. The quantitative standards would be formulated to address prudential concerns, and to ensure that the dispersion between the results of different proprietary models for a uniform set of market positions is confined to a relatively narrow range. They would be expressed in terms of a series of broad risk measurement parameters for banks' internal models, together with rules for converting the in-house measures of exposure into a supervisory capital requirement. The qualitative standards would ensure that the institutions' risk measurement systems are conceptually sound and that the risk management process is carried out with sufficient integrity. Additionally, to implement this framework it would be necessary to prepare harmonized guidelines on supervision and validation procedures for the private-sector monitoring agents, external auditors and regulatory agencies who would be charged with independently reviewing and certifying the institutions' internal systems.

In contrast to a pure standards-based framework, such as the CAD, this approach to financial market regulation, which this chapter terms the market surveillance approach, operates indirectly, through the deterrent effects of a decline in market value and associated loss of credibility which is suffered by an institution perceived to be acting imprudently. When operating effectively, it ensures the imposition of penalties on an aberrant institution which are in proportion to the damage caused.

The chapter addresses a number of issues. What are the major factors determining the relative desirability of standards and market surveillance as approaches to financial market regulation? What is the socially desirable way to employ the two approaches for dealing with market risk? Which approach to financial market regulation can best accommodate the dual stated objectives of the European Union's regulatory regime, namely ensuring the prudential safety of the financial system while simultaneously removing regulatory-induced competitive distortions in the market for financial services? What are the implications of the approaches for the regulation of offshore banking and financial activity in islands and small states?

In addressing these questions this chapter adopts an unashamedly instrumentalist method of analysis. The effects of standards and market surveillance are compared and then evaluated on a utilitarian basis. We work on the following assumptions: (a) that decision-making within each individual economic institution is directed towards maximizing that institution's market value; and (b) the regulator, for a given set of regulatory objectives, selects that regulatory regime which minimizes the sum of financial institutions' expenditures on risk management and the expected costs of systemic risk damage to the financial system. In the framework we adopt, this is equivalent to net social cost minimization.

The chapter's development is easily summarized. Using a model developed by Shavell (1984a), we outline the basic theoretical determinants of the relative desirability of capital standards and market surveillance. We then develop a procedure for incorporating these determinants into the analysis of financial market regulation which explicitly recognizes the heterogeneity of financial institutions. Capital standards and market surveillance mechanisms (which incorporate proprietary in-house systems) are then compared as mechanisms for alleviating the potential damage arising from systemic financial crisis. In a subsequent section we examine the conditions under which capital standards and market surveillance can be optimally combined in the design of an efficient regulatory regime. Following that, we discuss the implications of the results for the regulation of offshore banking and financial services activity, and close the chapter with some concluding remarks.

Determinants of the Relative Desirability of Capital Standards and Market Surveillance

To assess the determinants of the relative desirability of capital standards and market surveillance it is necessary to specify our criterion for comparison. Here, this criterion is assumed to be a desire to minimize the expected (net) social costs associated with abiding by a given set of regulatory constraints. These costs include both those associated with the occurrence of a systemic crisis in financial markets and those incurred by institutional adherence to any resulting regulations.

We are now in a position to outline the major theoretical determinants that influence the solution to the appropriate choice of regulatory regime by application and extension of a framework of analysis originally developed by Shavell (1984a). These determinants are as follows. The first is the extent of the difference in knowledge as between financial institutions and a regulator concerning the nature and effect of activities which increase market risk. These differences could relate to the benefits of the activities, the costs of institutional risk management, or the probability and/or extent of the risks faced. Second is the level of the financial institution's liquid assets relative to the potential systemic damages accruing to the financial system from the institution's risk-augmenting activity. Third is the possibility that any systemic damage occurring would not be correctly attributed to the originating institution. Finally, there are the administrative costs of implementing the different regulatory regimes. We now briefly consider each of these in turn.

Differences in information relating to risky activities

All other things being equal, and provided they possess the correct incentives to take adequate precautions, then if financial institutions have superior information relating to the benefits of risk-augmenting activities, the costs of institutional risk management, or the probability and/or extent of the risks faced, it is efficient to let the institution itself decide about the extent of risk management activity. For the sake of argument, consider the situation where financial institutions possess full information about activities enhancing market risk, whereas a regulator has incomplete information. To grant the regulator the power to impose restrictions on such activities in the form of a uniform set of standards is then likely to create a large probability of error. Underestimating the potential for systemic damage, or overestimating the contribution of the activity to the institution's profitability or the costs of risk management, will lead the regulator to establish a set of capital standards that are too lenient. The reverse mistakes will lead to standards that are too stringent.

Under market surveillance, the results may improve. This is obvious in situations where owing to the effective implementation of the technique of consolidation, institutions effectively bear full liability for their activities, suffering a loss in market value equal to the damage to the financial system they occasion as a result of their trading activities. Here, financial institutions are given the correct incentives to balance the true costs of risk management against the resulting loss of value to the institution arising from aberrant behaviour. These considerations, especially the use of consolidation, are particularly important when analysing the appropriate regulatory structure for offshore financial activity in islands and small states. We return to this issue later.

Two further questions remain. When can significant differences in information be expected to exist between financial institutions and the regulator, and, given regulatory objectives, when are such differences in information important? The first question can be handled fairly easily. Given that the financial institutions are engaging in, and deriving benefits from, the risky activities, one would expect that they would generally enjoy an inherent advantage in calculating the value of such benefits and the nature and costs of the resulting risks. Comparable information could generally be obtained by a regulator only from continuous observation of an institution's activities, a practical impossibility.

Of course, the familiar exceptions which may give a comparative advantage in information acquisition to a regulator are also possible. Information regarding the real benefits and costs of risk may not be revealed simply as a result of engaging in risky activities, but may require specialist expertise to evaluate. Under such circumstances a regulatory authority may be able to devote social resources to risk evaluation, whereas private parties may have insufficient incentives to do so if they are unable to capture the full benefits from so doing for all the familiar reasons associated with free-rider effects, and the inability to internalize fully the effects of their risk activities on third parties. Moreover, for individual institutions to undertake individually to acquire the same or essentially similar information may result in wasteful duplicative expenditures. However, even when a regulator acquires information, difficulties may be encountered in appropriately conveying that information to financial institutions because they are too numerous or difficult to identify.

Interestingly, informational criteria such as those considered above were the focus of a recent influential G30 report (Group of Thirty, 1993) which studied the risks emanating from the OTC derivatives market. Although this report does not examine issues relating to the regulation of market risks in any great detail (the issue of capital adequacy is not addressed), it applauds the high standards of risk control within the financial services industry, commenting 'the risks associated with complexity, concentration, liquidity and linkages between markets are manageable and being managed'. The tenet of the report is that superior informational processing capabilities are sufficient to enable the financial system to rely on market participants themselves to regulate their activities.

Incentive dilution and the level of the financial institutions' assets

The view expressed above is surely too simplistic. Given the nature of systemic financial crises and the real possibility of generalized financial contagion, it is very plausible that market surveillance would not provide adequate incentives to manage risk. Financial institutions are likely to treat any systemic damage caused to the financial system that exceeds the value of their assets as imposing liabilities equal only to the value of those assets. Under the imposition of capital standards, any inability to pay for harm caused is irrelevant, assuming that all institutions are compelled to fulfil the requisite standards as a pre-condition for engaging in their activities. In assessing the force of this argument, which favours capital standards over market surveillance, one factor which must be considered is the extent of the regulated financial institution's assets relative to the probability distribution of the magnitude of systemic damage to the financial system that may be occasioned by imprudent behaviour. The smaller the former relative to the latter, the greater the appeal of capital standards.

At this point, we raise an additional consideration which is relevant when assessing the regulatory implications of incentive dilution. This concerns the impact of delegating decision-making to individual employees within financial institutions.[5] As events during 1995 at institutions such as Barings and Daiwa Bank clearly illustrate, decisions made by employees can result in damage both to a specific institution and to the financial system as a whole, far in excess of an individual employee's assets. Although this gives institutions good reasons to establish appropriate proprietary controls over their employees' behaviour, the situation becomes more complex upon recognition that: (a) the requisite knowledge of those individuals entrusted with implementing the internal controls may be lacking, and (b) the assets of these internal supervisors are also likely to be less than the potential liabilities arising from the breakdown of internal control mechanisms. Under such circumstances, there exist other arguments in favour of establishing mandatory capital standards which are independent of the dilution incentives arising from comparison of the level of a financial institution's assets in relation to its potential liabilities.

Incentive dilution and the correct attribution of damages

In a similar vein to the previous determinant, the possibility that the source of systemic financial crisis may not be correctly identified or is incorrectly attributed will also result in a dilution of the incentives to engage in the appropriate level of risk management activity under a market surveillance regime. The importance attached to this factor depends to some extent upon the reasons why incorrect attribution may occur. However, what is clear is that a necessary condition for market surveillance to provide the correct incentives for financial institutions to manage risk efficiently is that consolidated supervision be managed effectively. This observation reinforces the primary principle to emerge from the 1983 Basle Concordat, namely that the soundness of a bank cannot be properly assessed unless supervisors or regulators can examine the totality of each institution's business worldwide. In other words, consolidated supervision must be implemented effectively. This point is particularly important in the context of the regulation of banking activity in islands and small states, as the BCCI case clearly demonstrates. Moreover, the difficulties of correct attribution may only be compounded when the source of the damage is traced to decision-makers within complex multinational financial institutions, or where fraudulent institutions deliberately set out to confuse regulators. Even within a given institution it may be impossible to allocate individual responsibility correctly, or those individuals deemed responsible for the damages caused may no longer be subject to effective discipline. We expand on these themes later in the chapter.

Costs of alternative regulatory regimes

The final determinant we consider is the size of the administrative costs of the regulatory system. These are broadly defined to include all the public expenditure associated with litigation, settlement, the maintenance of the regulatory establishment(s), and the private-sector costs of establishing risk management systems to comply with the regulatory regime which is imposed. Here, little definitive can be said without reference to details of the specific system in place. A couple of points are, however, worth noting if we are allowed to make the heroic assumption that the total risk management costs incurred by financial institutions are the same under both regimes. These both tend to favour the market surveillance approach. One cost advantage of the market surveillance approach is that it naturally directs resources to the control of activities most likely to lead to systemic damage, for under this framework institutions should be induced to engage in the appropriate level of risk management for each risky activity. Under capital standards, in the absence of specialist information about different institutions' potential contribution to risk, there is no tendency for administrative costs to be focused on those activities most likely to cause systemic damage within specific institutions. Moreover, if the market surveillance approach is functioning correctly, in the extreme case all institutions will undertake the appropriate level of risk management, all litigation and settlement costs will be mitigated, and, other than certain fixed costs, all administrative costs will be reduced to zero.

A Framework for Analysing Financial Market Regulation

Heterogeneity of financial institutions and regulatory policy

Any attempt to integrate the previously outlined theoretical determinants into a general analysis of financial market regulation must begin by acknowledging a central stylized fact. This is the fundamental heterogeneity of institutional business characteristics which exist within the financial sector, particularly across the division between the banking and securities industries. The consensus that considerations relating to both institution-specific risks and the composition of assets and liabilities imply that bank insolvency is potentially a greater contributor to systemic risk damage within the financial system than the insolvency of a typical securities firm is adequately illustrated by Dale (1992a, b) and Steil (1994a). Specifically, banks' main assets are primarily illiquid commercial loans held on the balance sheet until maturity. The fact that this large component of their asset base is not marketable implies that banks in distress cannot be liquidated quickly without jeopardizing their creditors' claims. The situation is exacerbated by the fact that a major component of a bank's non-capital liabilities are retail deposits which can be withdrawn on demand. The associated risks make banks uniquely vulnerable to contagious deposit withdrawals, with consequent implications for the prudential safety of the entire financial system.

In contrast, securities firms experience rapid turnover of their asset bases as a result of their market-making, trading and underwriting activities. Moreover, these assets are, by definition, highly marketable. Thus, securities firms can adjust their balance sheets and their market risk profiles rapidly in response to changing circumstances. On the liabilities side, while securities firms are reliant upon the wholesale money markets for their non-capital funding, a large proportion of that borrowing is secured, and therefore unlikely to be subject to sudden contraction.

It follows that regulatory capital requirements must serve different functions for banks and securities firms. With respect to banks' capital requirements, the focus must be directed towards permanent capital, intended to support and sustain the continued financial viability of any troubled institution. In contrast, securities firms encountering financial difficulties should be expected to contract their balance sheets accordingly,[6] which requires an emphasis on temporary or liquid capital. This explains securities firms' ability to rely heavily upon short-term subordinated debt as a source of capital.[7] Banks can generally include such financing only in 'secondary' capital, subject to strict limits in terms both of its amount and its term to maturity.[8]

These factors are important in view of the recent advocacy of activity-based regulation, and the consequent desire to impose uniform capital standards to govern the composition of the trading books of both banks and securities firms. The rationale for this development is the view that two institutions engaged in an identical activity should, with respect to that activity, be subject to identical regulatory standards. However, for the purpose of quantifying an institution's potential contribution to systemic risk damage to the financial system as a whole, its trading book cannot be viewed in isolation from that institution's other financial functions. Furthermore, although trading book performance influences an institution's ability to perform such functions, these functions differ between banks and securities firms. All these considerations suggest not only that it is inappropriate to segment an institution's financial activities for the purposes of calculating capital requirements, but also that the optimal requirements are likely to differ between banks and securities firms.

Dale (1994) provides an excellent critical discussion of all of these issues, from an informal perspective. In the following section we trace through the implications of institutional heterogeneity for the impact of the previously outlined theoretical determinants of the optimum regulatory structure.

Determination of optimal capital standards

The optimal capital standard under the prudential safety objective

Consider a group of heterogeneous financial institutions consisting of banks and securities firms, which can reduce the probability of a systemic crisis in financial markets arising as a result of their securities trading by undertaking costly risk management activities. This risk management includes implementing the provisions of regulatory directives (such as the CAD), satisfying mandatory disclosure requirements, and instituting proprietary in-house risk management systems.

Initially, we focus on a situation where a risk-neutral financial market regulator has been entrusted with the sole objective of maintaining the prudential safety of the financial system. To do this, the regulator must specify a CAD-type directive incorporating a set of risk-based capital requirements. The requirements are defined in terms of their cost of implementation. To begin, we rule out differences in information between financial institutions and the regulator, and assume that all parties possess full information on each specific institution's potential contribution to systemic risk.

As institutions differ in their capacity to contribute to systemic risk damage, the regulator should impose a different set of capital standards upon each institution, specifically one which equates that institution's marginal cost of risk management to the expected marginal benefit of the standard in reducing the social cost of financial crisis.[9] We assume that increased expenditure on risk management by an institution reduces the possibility of its contributing to a systemic crisis, but that there are decreasing returns to such expenditure. It follows that the optimal extent of risk management, by an institution, is increasing in that institution's potential contribution to systemic risk. This is illustrated by X^*X^* in Figure 13.1.

The optimal capital standard under the level-playing-field objective

We now analyse the position of a regulator charged solely with satisfying the level-playing-field objective. If all institutions are assumed to face the same costs of risk management for a particular activity, regulatory neutrality implies that any activity-based regulatory standards, for example the standardized set of capital requirements required under the CAD, must be the

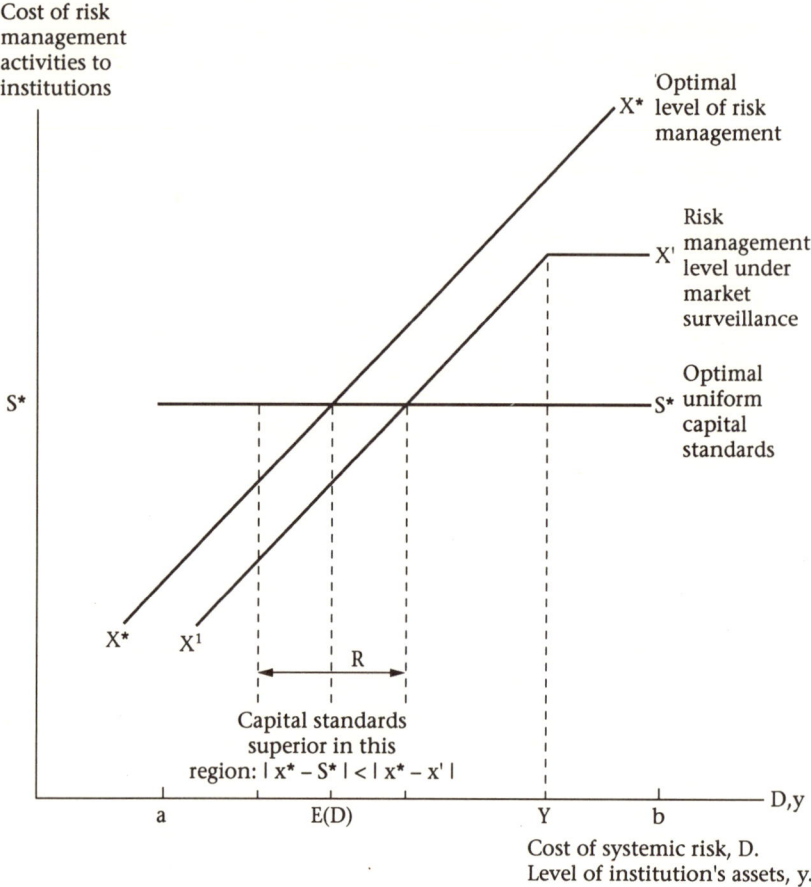

Cost of risk
management
activities to
institutions

'Optimal
X* level of risk
management

Risk
X' management
level under
market
surveillance

S*

Optimal
S* uniform
capital
standards

X*   X¹

R

Capital standards
superior in this
region: | x* – S* | < | x* – x' |

D,y

a   E(D)   Y   b

Cost of systemic risk, D.
Level of institution's assets, y.

Figure 13.1   A comparison of risk management activity under uniform capital standards and
market surveillance

same for all financial institutions. Given diminishing returns to risk management expenditure, the optimal set of capital requirements under the
level-playing-field objective will correspond to those that are efficient from
the prudential safety perspective only for an institution expected to impose
the average level of systemic risk damage on the system. The optimal
uniform standards are denoted S*S* in Figure 13.1

The conflicting nature of the two regulatory objectives under a CAD
uniform standard approach is now readily apparent. For institutions with the
potential to create above (below) average levels of systemic damage, the set
of requirements designed to level the playing field will be set too low (high)
from the prudential safety perspective. Moreover, as the CAD incorporates a
single set of activity-based capital standards to be applied to the trading
books of all financial institutions, the foregoing discussion suggests that the

interests of prudential safety have been subsumed to those of promoting competitive equality during the regulatory design process.

The difficulty of reconciling these dual objectives has already become apparent. In April 1993 the Basle Committee on Banking Supervision published a set of consultative proposals for the regulatory treatment of market risks incurred by banks, which covered trading book positions in certain derivatives, debt securities, equities, commodities and foreign exchange. The approach taken by this initial proposal (Basle Committee, 1993) closely resembled that of the CAD in the manner in which it measured and assessed market risk as a basis for the imposition of activity-based capital charges. Indeed, in an explicit attempt to secure equality of regulatory treatment for banks and securities firms, the Committee proposed that banks should be permitted to employ an additional form of short-term subordinated debt for the sole purpose of meeting part of the capital requirement for market risk. This was a feature of the proposals which 'bank supervisors acting on their own would not favour but are prepared to adopt in the hope that further convergence with securities regulators will be achieved at some future date' (Basle Committee, 1993a, p. 2).

Notwithstanding the fact that, in this way, a degree of regulatory neutrality is secured between banks and securities firms, this reaction of bank supervisors explicitly recognizes the difficulties that institutional heterogeneity poses for joint satisfaction of the prudential safety and level-playing-field objectives under any standards-based approach, such as the CAD.[10]

A reconciliation of regulatory objectives invoking incomplete information

We now relax the assumption of full information on the part of the regulator. As mentioned earlier, given that the financial institutions are engaging in, and deriving benefits from, the risky activities, it is to be expected that they would generally enjoy an inherent advantage in calculating the value of such benefits, the nature of risks created, and the costs of their reduction.

Certain consequences of relaxing the full-information requirement are immediately apparent. In order to establish the socially efficient capital standards under the prudential safety objective, the regulatory authority requires full information concerning each specific institution's potential contribution to systemic risk damage. The informational requirements for setting the optimal capital standard under the level-playing-field objective are much weaker. Specifically, it is inconsequential whether the regulator possesses complete information for each individual institution, or only has information on the expected extent of the systemic damage across the financial system. The capital standards set would be the same in both cases. Moreover, if the latter is the extent of the regulator's information, then even from a prudential safety perspective, the imposition of a uniform set of capital standards can also be justified as a second-best response to this particular form of incomplete information on the part of the regulator.

Market surveillance approach to financial market regulation

So far we have discussed the issue of financial market regulation as if the supervisory function were the exclusive domain of a regulatory authority. We now turn our attention to the alternative, market surveillance view, which holds that the emphasis in financial market regulation should be directed towards developing more efficient proprietary, in-house practices for managing market risk, supplemented with consolidated, market-based scrutiny where required. The market surveillance approach operates indirectly, through the deterrent effects of a decline in market value and the associated loss of credibility which is suffered by an institution perceived to be acting imprudently. When operating effectively, it ensures the imposition of penalties on an aberrant institution which are in proportion to the damage caused.

The qualifications usually attached to proposals to implement a market surveillance approach to regulation are those identified on p. 248. They arise from two main considerations, which both operate via the dilution of an institution's incentives to undertake the appropriate level of risk management. First, the cost of the damage arising from an institution's contribution to systemic risk may exceed the level of the institution's assets. Second, given the problems which arise from interpreting informational signals, the difficulties of effectively implementing consolidated supervision, and the potential difficulties of correctly attributing responsibility for systemic damage once it has occurred, there is the possibility that imprudent behaviour will go undetected, or that the institution responsible will not be held fully accountable for the damage it causes.

If an institution is deemed to be fully accountable for the level of systemic damage it causes under a market surveillance regime, then the imposition of full liability ensures that the value of the institution's assets will fall by an equivalent amount. Clearly, the institution will bear the full cost of its actions if and only if the damage it causes is no greater than the value of its assets.

Knowing this, the institution will choose its level of proprietary risk management to minimize (a) the cost of such activity, plus (b) the expected loss in its asset value attributable to any penalties it is expected to suffer as a result of its contribution to systemic risk damage, given its chosen level of risk management. So long as there is some probability that the financial institution will not be held fully accountable for the contribution of its activities to systemic risk, or the level of its assets will be insufficient to cover its liabilities, then the level of risk management selected by the institution will be less than that which is socially efficient, from the prudential safety perspective, but will increase with the level of damages as long as the damage it causes is no greater than the value of its assets. Thereafter it will be constant. This is illustrated by X'X' in Figure 13.1. Note that as a market surveillance regime delegates responsibility for risk management to the financial institutions themselves, considerations relating to informational asymmetries are not relevant to the determination of the level of risk management that will be undertaken. Moreover, as supervisory standards vary considerably between financial centres, especially offshore centres, reliance on market surveillance renders impotent the regulatory objective of levelling the playing-field. We return to discuss further implications of this issue later in the chapter.

Comparison of capital standards and market surveillance

We can now compare capital requirements and market surveillance as mechanisms for alleviating the potential damages arising from systemic risk. We adopt as our criterion for comparison the difference in expected social costs between the two regulatory regimes, one which employs market surveillance alone and the other uniform capital requirements alone. In the former case we assume that consolidated supervision is effective. In the latter case we assume that the regulator imposes the optimal set of uniform standards in accordance with the stipulated objective of levelling the playing-field.

It turns out, perhaps not surprisingly, that the imposition of capital standards is superior to market surveillance if the majority of the costs associated with systemic risk damage are expected to lie sufficiently close to the average expected level of damages. In terms of Figure 13.1, this means the majority of costs lie in the region R, defined as the area where the absolute value of the distance between X*X* and S*S* is less than that between X*X* and X'X'. This situation is most likely to arise if either (a) the factors which reduce the incentive to manage market risk under market surveillance are sufficiently important, or (b) the variability among parties in their contribution to systemic risk damage is sufficiently small. The less heterogeneous the financial institutions in terms of their trading activities, the more likely it is that uniform capital standards will dominate market surveillance.

Alternatively stated, whenever there exists a wide disparity in the ability of financial institutions to contribute to systemic risk, the desire to promote a 'level playing-field' as a regulatory objective through activity-based regulation may well conflict with considerations of regulatory efficiency, defined in terms of the desire to minimize the sum of the social cost of the regulatory constraints and the systemic damage their imposition is designed to preclude. In particular, there are grounds for arguing that the implicit belief underlying the imposition of uniform capital standards in the CAD, namely that allowing securities firms to compete on more favourable regulatory terms than banks, or vice versa, would be both inequitable and damaging to the efficient operation of financial markets, is generally misplaced.

A Regime Combining Capital Standards and Market Surveillance

Following further consultation with financial institutions which lasted from April to December 1993, the Committee revised its proposals (Basle Committee, 1995) to address the following criticisms. First, the initial proposals did not recognize best market practice in risk measurement techniques, and thus failed to provided a sufficiently strong incentive for institutions to improve risk management systems. Second, the proposed risk assessment methodology failed to take account of correlations and portfolio effects across all instruments and markets, and generally did not sufficiently reward risk diversification.[11] Third, the initial proposals were insufficiently compatible with the banks' own, often far more sophisticated, measurement systems, implying unnecessarily high compliance costs.

The amended proposals, analysed by Hall (1995b), sanction the use of proprietary models for the calculation of market risk as an alternative to the standardized measurement framework, provided certain conditions are fulfilled. These conditions relate to the following: (a) general criteria concerning the adequacy of the risk management system; (b) qualitative standards for internal oversight of the use of models, notably by management; (c) guidelines for specifying an appropriate set of market risk factors; (d) quantitative standards setting out the use of common minimum statistical parameters for measuring risk; (e) validation procedures for external oversight of the use of models; and (f) rules for banks which use a mixture of models and the standardized approach. All elements of market risk which are not captured by an internal model, or which are relevant for those banks not using internal models, will remain subject to the standardized measurement framework.

To analyse this approach, consider an environment in which institutions must satisfy a set of capital standards and/or fulfil approved market surveillance procedures. In order to respond efficiently to the regulations, institutions will undertake a level of risk management which is at least equal to that mandated by whatever regulatory standard is established, say $S'S'$. They will also exceed this if it is in their own self-interest to do so, i.e. if $X'X' > S'S'$.

The resulting extent of risk management in financial markets will be that depicted in Figure 13.2. For a set of standards such as S_1, financial institutions for which the standards are too strict from an efficiency perspective will devote resources to managing risk to the level S_1, and others, for which they are too lax, will manage risk to the level $X'X'$. In contrast, for requirements

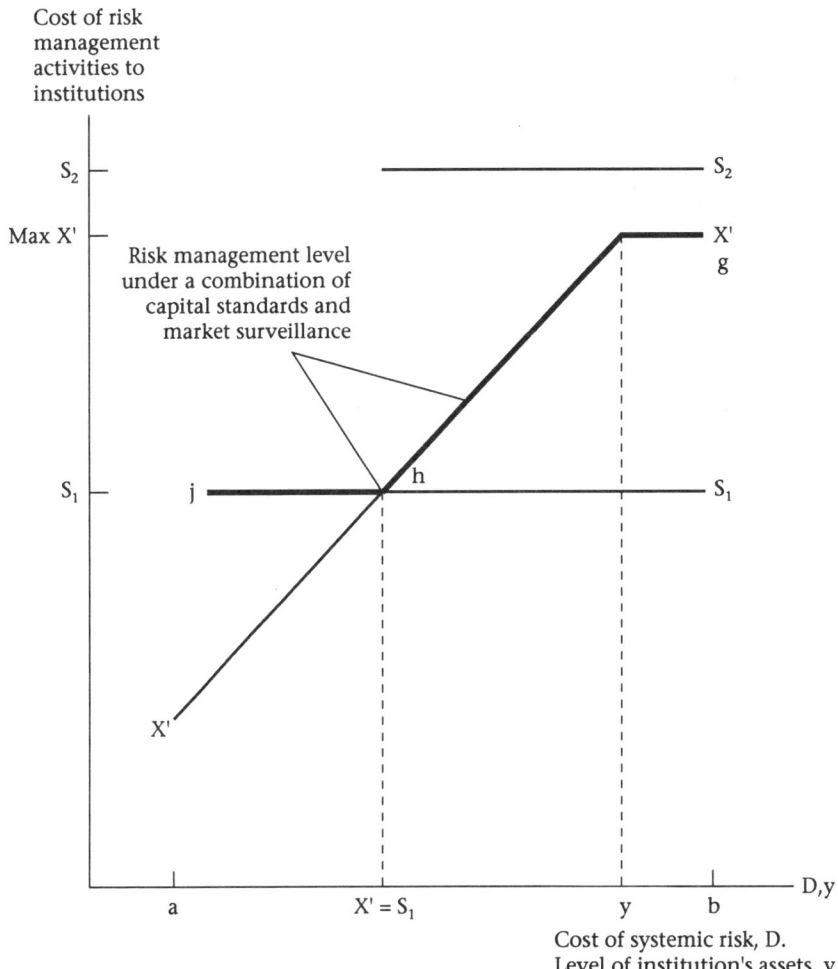

Figure 13.2   Risk management levels under a combination of capital standards and market surveillance

such as S_2 that exceed $X'X'$ for all levels of expected damage, all institutions will manage risk to the level S_2.

Being aware of the response of institutions to the regulations, the problem for the regulatory authority is to set the external capital standards at the level where the marginal cost to an institution of increasing the level of its risk management activity is equated to the expected marginal reduction in systemic risk costs that this enhanced activity brings about. Here, it should be noted that the expectation is taken only over those institutions that remain unaffected by market surveillance penalties and, therefore, are required to adhere to the mandatory capital standards.

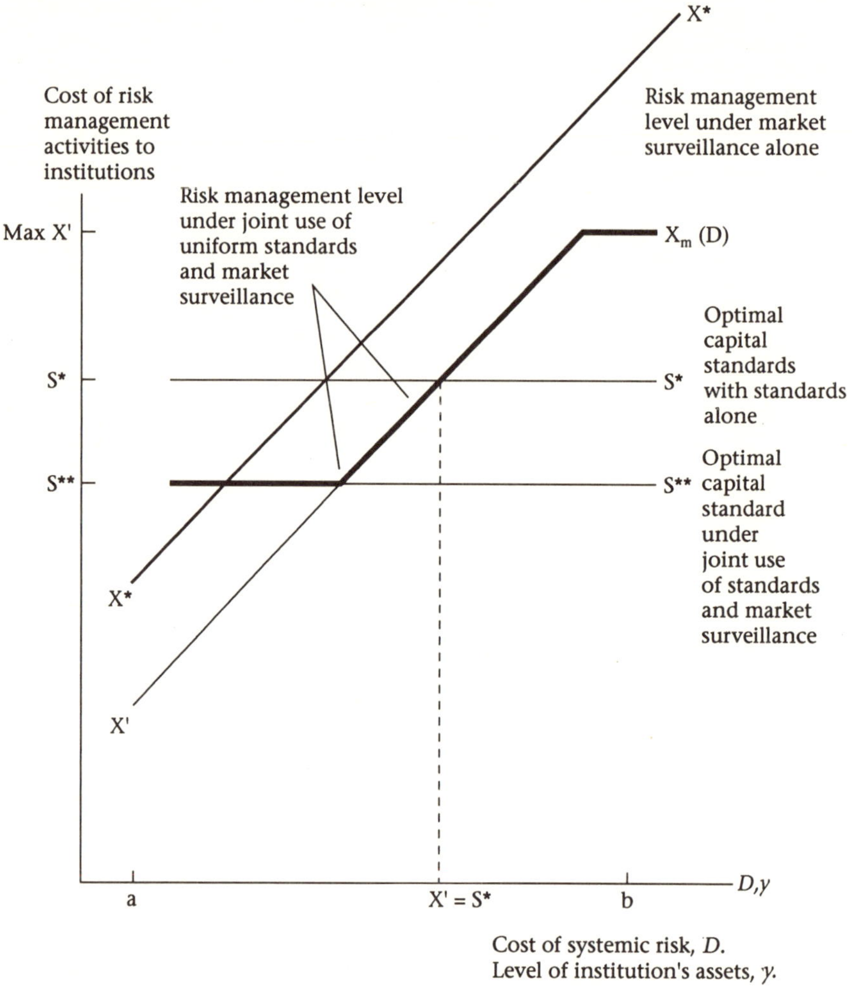

Figure 13.3   The optimal joint regulatory standards: limited incentive dilution

Two general classes of outcome exist. In the first class, illustrated in Figure 13.3, the optimal regulatory standard, S**, under the combined use of standards and market surveillance (characteristic of the revised 1995 Basle Committee approach), is less than the optimal standard, S*, if capital requirements are used alone (as under the CAD). However, the combined standard, S**, still exceeds the first best level of risk management for those institutions least likely to pose systemic risk problems.

The optimal combined regulatory standards, S**, will be exceeded by certain institutions. They are induced to exceed them by the fact that the resulting expected loss in the institution's value if its risk control practices are revealed to be suboptimal exceeds the institution's costs of implementing the appropriate risk control procedures. The incentive therefore exists to enact those procedures. A sufficient condition for this to occur is that the institution's incentive to manage risk is not excessively diluted, because of the considerations discussed earlier.

The reasons why S** < S* are worth clarifying. Essentially, when capital standards alone are used, reducing the set standards below S* is not worthwhile from a social perspective, as it results in all parties undertaking less risk management than S*. When capital standards are combined with effective market surveillance, only those parties with a lower than expected contribution to systemic damages undertake less risk management activity. The 'riskier' institutions are induced by market surveillance to manage risk to a

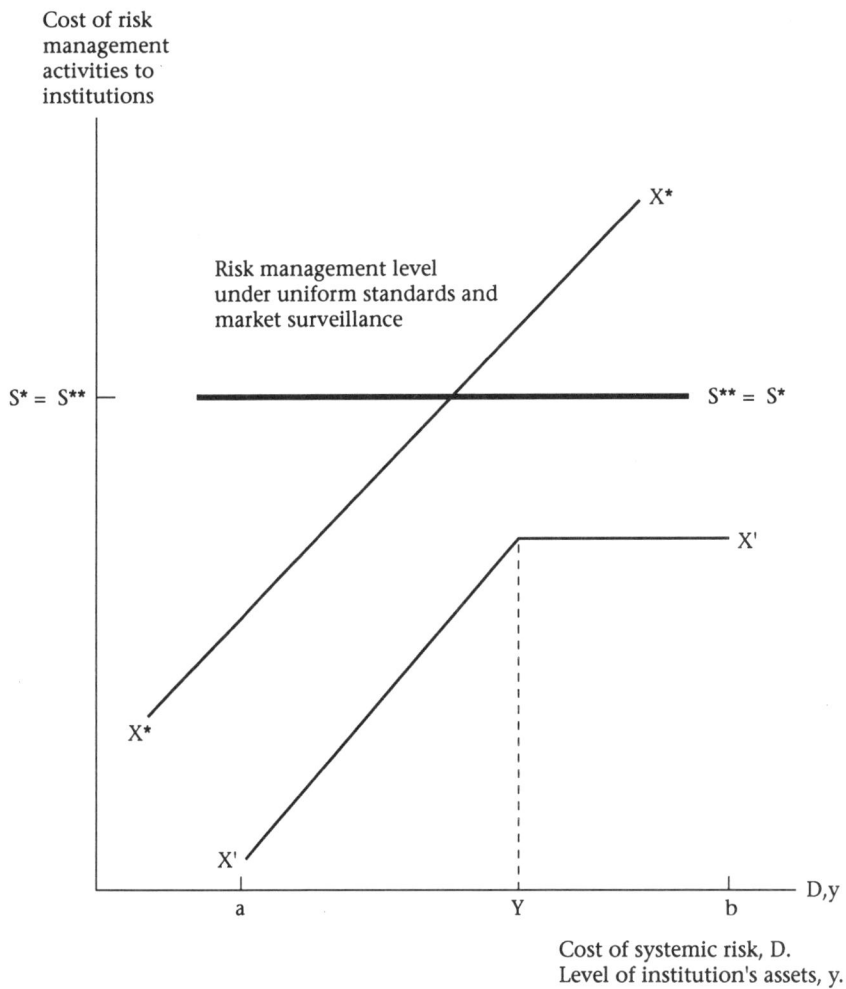

Figure 13.4 Optimal joint standards: severe incentive dilution

level greater than required by simple adherence to the requisite capital standards, S*. Such action augments the efficiency of the financial markets.

To understand why the capital standards exceed the level of risk management that would be efficient for an institution whose trading activities pose the least threat of contributing to financial crisis, note that raising the requisite standard from this minimum level does not lead to any significant change in the expected social costs associated with systemic risk crisis (and its prevention) for these institutions. However, it does result in a significant reduction in the expected social costs associated with systemic damage for all those institutions that would not be induced by market surveillance alone to undertake risk management activity at a level equal to the minimum socially desirable level.

The second class of outcomes, illustrated in Figure 13.4, arises when no institution can be induced by market surveillance penalties alone to devote resources to risk management to a greater extent than that required by the externally imposed capital standards. This situation occurs if the incentive to manage risk in-house is so sufficiently diluted that proprietary risk management mechanisms are unable to take up any of the regulatory slack from lowering the external capital standards below S*. In this case, the level of the optimal combined capital standards equals that when standards alone are employed.

Implications for the Regulation of Offshore Financial Activity

We now turn to consider the implications of the preceding analysis for the regulation of banking and financial activity in islands and small states. We remind the reader that banks in offshore centres (and elsewhere) are subject to comprehensive regulatory oversight, because history has demonstrated that left to their own devices, *some* institutions will incur risks, market-linked or otherwise, that are excessive when judged by the damage which their failure can inflict on the financial system. Moreover, if a major portion of the offshore financial industries' business becomes too complex and opaque to regulate efficiently, certain institutions will undoubtedly try to exploit this situation to escape from the regulatory constraints on risk-taking, thereby attempting to augment their competitive position.

We now assess the supervisory weaknesses revealed by the collapse of the Bank of Credit and Commerce International (BCCI) in 1991. These are used as evidence in assessing the relative merits of the previously outlined regulatory approaches in the context of regulating offshore banking activity in islands and small states.

BCCI was established in 1972 by a group of bankers from Pakistan with capital contributions from Bank of America and Middle Eastern investors.

Bank of America sold its stake in 1980, leaving prominent Middle East-based financiers, notably the Bin Mahfouz family of Saudi Arabi and the family of Sheikh Zayed Bin-Nahyan, the ruler of Abu Dhabi, as the leading shareholders.

By 1990 BCCI was one of the largest private banks in the world, with stated assets of around $20 billion, and branches operating in some 70 countries. This global banking network was managed through an unusual corporate structure, involving a parent holding company in Luxembourg (BCCI Holdings), with the main banking business split between two subsidiaries incorporated in different offshore jurisdictions, namely in Luxembourg (BCCI SA, established in 1972) and the Cayman Islands (BCCI (Overseas), established in 1976).

On 5 July 1991, under a Bank of England-orchestrated operation, co-ordinated action was taken in more than 60 countries around the world to close down the activities of BCCI. Direct action was taken to seize the group's assets in eight countries initially – the United Kingdom, the Cayman Islands, the United States, France, Germany, Spain, Switzerland and Luxembourg – and the branches operating within those jurisdictions were closed. The grounds cited for this action comprised dishonest and fraudulent management of the bank; concealment of fraudulent activities and their consequences from the Bank of England and other regulatory authorities; concealment of the bank's true financial position; inadequate accounting records; inadequate management controls; managers' acting without integrity and with lack of skill; loss of trust and confidence in senior management of the bank on the part of the Bank of England; the insolvency of the bank.[12]

This chapter addresses two issues emanating from the BCCI affair which are of direct relevance to the preceding discussions. First, a necessary condition for the market surveillance approach to work effectively is that in order for the soundness of a bank to be assessed, regulators must be able to examine the totality of a bank's worldwide business through the technique of consolidation. BCCI was able to use its unusual corporate structure to evade effective consolidated supervision of its worldwide operations and to shield its activities from prying eyes by hiding under tough bank secrecy laws prevailing in certain offshore centres,[13] and to confuse regulators deliberately by shuffling its assets between different jurisdictions.

This was achieved by establishing a holding company in Luxembourg which, like its notorious forerunner, Banco Ambrosiano Holdings, escaped detailed supervision by the Luxembourg authorities, and then setting up subsidiaries in Luxembourg and the Cayman Islands. Branches of the former subsidiary were used to engage in banking operations in the United Kingdom, where the bank's operational headquarters was situated until it was moved, at the request of the Bank of England, to Abu Dhabi, the domicile of the major shareholders.

Somewhat belatedly (in 1987), a college of regulators was set up (initially comprising the supervisory authorities of the United Kingdom, Switzerland, Spain, Luxembourg, France, the United Arab Emirates, Hong Kong and the Cayman Islands), to ensure effective and coordinated supervision of the BCCI group's activities. The dispersal of supervisory responsibilities within the group and the difficulties in getting agreement on coordinated action, however, served only to delay the day of reckoning for BCCI, although some positive steps (e.g. in securing agreement to a restructuring and recapitalization of the banking operations and to a replacement of senior officers) were taken.

The fact remains, though, that, despite the 1983 revision of the Basle Concordat, which governs the division of supervisory responsibilities between parent and host bank supervisors, the agreement clearly failed to ensure that BCCI was effectively regulated on a consolidated basis. As effective implementation of consolidated supervision is a necessary condition for a market surveillance approach to operate efficiently, the BCCI case illustrates the difficulties of implementing such an approach in offshore financial centres.

Second, the BCCI case demonstrates the ease with which a fraudulent bank can exploit weakly regulated offshore centres in the absence of any machinery for ensuring the adequacy of supervisory standards. On this issue the Bingham Report on BCCI concludes that 'the need for some independent monitoring of supervisory standards is in my view clear', and that host supervisors 'must be reassured by some form of independent verification that the home supervisor is really doing his job' (Bingham, 1992, at 3.27).

The Basle Committee attempted to meet these concerns by introducing the requirement that all banks should be supervised by a home-country authority 'that capably performs consolidated supervision'. However, the effectiveness of these new guidelines depends on the ability of national authorities to monitor each other's quality of supervision, so that they can exercise an informed judgement as to whether banks from a particular jurisdiction should be excluded from their territory or subject to special restrictions imposed by the host authority.

Under the Basle guidelines one country is called upon to assess another country's quality of supervision on the basis of the latter's statutory powers, administrative practices and supervisory records. But there is no new multilateral machinery to assist in the monitoring process, and it is difficult to see how bilateral relationships can provide adequate information about supervisory standards in particular jurisdictions. This concern is magnified in the case of offshore financial centres in islands and small states. In order to overcome this problem, the Bank of England has proposed a system of peer group review under which each country's supervisory arrangements would be assessed by a panel of supervisory authorities from other countries. However, it is as yet unclear whether this kind of approach would have the broad support among offshore centres that would be needed to make it viable. If the Basle approach proves inadequate to the task, then there may eventually be pressure for a more formal and detailed approach, together with concentrated international action to isolate those offshore centres that make a virtue out of their bank secrecy laws.

Conclusions

This chapter analyses the theoretical determinants of the relative desirability of capital standards and market surveillance as approaches to financial market regulation. The European Commission's Capital Adequacy Directive (CAD), which governs the regulation of market risk in the European market for financial services as from 1 January 1996, adopts a standardized measurement framework to establish a single set of minimum capital standards which are to be applied to the trading books of all financial institutions. In contrast, the Basle Committee on Banking Supervision has advocated supplementing a CAD-type standardized risk measurement framework with proprietary, in-house risk management procedures accompanied by market surveillance.

The chapter's main conclusions are as follows. The CAD's dual stated objectives of promoting competitive equality in the market for financial services while simultaneously ensuring its prudential safety are difficult to reconcile within a capital standards regulatory framework without recourse to *ad hoc* assumptions concerning the information set of the regulator. Moreover, our analysis indicates that the interests of prudential safety have been subsumed to those of promoting competitive equality in the regulatory design process underlying the CAD. We also analyse the circumstances under which regulatory efficiency in financial markets, including offshore centres, can be enhanced by supplementing a CAD-type standardized risk measurement framework with proprietary, in-house risk management procedures accompanied by market surveillance. One variation of this approach was recently proposed by the Basle Committee (1995). Here we conclude that, in general, it is socially efficient to implement a regulatory framework which combines the two approaches to regulation provided the requirements of consolidated supervision are effectively implemented. This suggests that the Basle Committee's approach may be efficiency-enhancing relative to that proposed under the CAD. The main exception to this finding arises when the level of an institution's assets are deemed insufficient to generate the correct incentives for that institution to introduce proprietary in-house risk management practices. This is particularly likely to occur if consolidated supervision cannot be implemented effectively. As the BCCI case illustrates, this presents particular difficulties in using the market surveillance framework to regulate financial activity in offshore centres.

Of course, the choices financial market regulators actually make between a regulatory framework employing market surveillance and one simply based on uniform capital standards are usually influenced by other factors lying outside the considerations discussed in this analysis. In any event, this choice will often not reflect a considered use of utilitarian, social cost–benefit analysis, such as that adopted in this chapter. Furthermore, the complexity of the relationship between standards and market surveillance also affords ready explanations of the varying approaches to the problem of market risk which have now been adopted by different regulatory agencies.

Notes

1. A credit institution is defined as an 'undertaking whose business it is to receive deposits or other repayable funds from the public and to grant credits for its own account' (see EC, 1977, art. 1; EC, 1989c, arts. 1(1) and 2(2)). The key characteristic of a credit institution is deposit-taking. The 'Second Banking Co-ordination Directive' (EC, 1989c), art. 3, provides that member states shall prohibit institutions that are not credit institutions from taking deposits from the public.
2. Investment firms are entities whose business it is to provide any investment service, defined as brokerage, dealing as a principal, market-making, professional investment advice, arranging or offering underwriting services, and safekeeping and administration of certain investments. See 'Investment Services Directive' (EC, 1993b), art. 1(2), (3); Annexe.
3. Specifically, the 'Own Funds Directive' (EC, 1989a) and the 'Solvency Ratio Directive' (EC, 1989b), which both took effect from the end of December 1992. With several minor differences, these EU directives largely converge with the requirements of the Bank for International Settlements (BIS) on capital adequacy. For more extensive discussion, see Gruson and Feuring (1991), Fowle (1991) and Hall (1989).
4. See EC (1993a). Hall (1995a) provides an assessment.
5. A detailed discussion of this issue is beyond the scope of the present study.
6. For example, the Securities and Exchange Commission's capital adequacy rules for broker-dealers in the US are specifically designed so that such firms can wind down their activities within one month. In the UK, the targeted wind-down period for troubled securities firms under the Securities and Futures Authority's regulatory framework is three months.
7. Under SEC rules for broker-dealers, a large underwriting may be capitalized with temporary subordinated debt repayable within 45 days.
8. See p. 257 for the CAD provisions on subordinated debt.
9. A formal presentation of this and the following propositions is given by Bowe and Hall (1996), an application of the Shavell (1984b) model to the analysis of regulatory regimes in financial markets.
10. Moreover, as many other commentators have explained, the amount of discretion which still resides with national regulators militates against the attainment of the level-playing-field objective, as does the availability of the banking single passport in the EU, which also extends to the provision of certain classes of investment services. Further detailed analysis of these and related issues is provided by Dale (1992a), Baltensperger and Dermine (1990), and the essays in Steil (1994b).

11. Both the original Basle Committee consultative proposal and the CAD promote the use of the building-block approach to risk assessment by splitting the total-position risk into specific and general risk components. Dimson and Marsh (1995) demonstrate the inefficiency of such an approach *vis-à-vis* a portfolio approach to risk assessment, a criticism the Basle Committee has acted upon in formulating its revised 1995 proposals (although not if banks adopt the 'standardized' approach).
12. See Hall (1991) for more detailed discussion.
13. For further discussion and details, see *The Banker* (1993), and the essays in Roberts (1994).

References

Baltensperger, E. and Dermine, J. (1990) 'European Banking: Prudential and Regulatory Issues.' In J. Dermine (ed.) *European Banking in the 1990s*. Oxford: Blackwell.

The Banker (1993) *Luxembourg: the Well-kept Secret*, October: pp. 33–34.

Basle Committee on Banking Supervision (1993) *The Supervisory Treatment of Market Risks: a Consultative Proposal*. Basle: Bank for International Settlements.

Basle Committee on Banking Supervision (1995) *Proposal to Issue a Supplement to the Basle Capital Accord to Cover Market Risks*. Basle: Bank for International Settlements.

Bingham, L. (1992) *Inquiry into the Supervision of the Bank of Credit and Commerce International*. London: HMSO.

Bowe, M. and Hall, M.J.B. (1996) *Rules with or without Discretion: A Comparison of Capital Standards and Market Surveillance as Mechanisms for Regulating Market Risk*. Loughborough: Loughborough University of Technology.

Dale, R. (1992a) *International Banking Deregulation*.' Oxford: Blackwell.

Dale, R. (1992b) 'Capital Adequacy and European Securities Markets.' In A. Steinherr (ed.) *The New European Financial Market Place*. London: Longman.

Dale, R. (1994) 'International Banking Regulation.' In B. Steil (ed.) *International Financial Market Regulation*. Chichester: John Wiley, ch. 7.

Dimson, E. and Marsh, P. (1995) 'Capital Requirements for Securities Firms.' *Journal of Finance*, vol. 50: pp. 821–851.

EC (1977) *Co-ordination of Laws, Regulations and Administrative Provisions Relating to the Taking-Up and Pursuit of the Business of Credit Institutions* ('First Banking Co-ordination Directive'), Council Directive 77/780/EEC, December.

EC (1989a) *Own Funds of Credit Institutions* ('Own Funds Directive'), Council Directive 89/229/EEC, April.

EC (1989b) *Solvency Ratio for Credit Institutions* ('Solvency Ratio Directive'), Council Directive 89/647/EEC, December.

EC (1989c) *Co-ordination of Laws, Regulations and Administrative Provisions Relating to the Taking-Up and Pursuit of the Business of Credit Institutions* (Amending Directive 77/780/EEC) ('Second Banking Co-ordination Directive'), Council Directive 89/646/EEC, December.

EC (1993a) *Capital Adequacy of Investment Firms and Credit Institutions* ('Capital Adequacy Directive'), Council Directive 93/6/EEC, March.

EC (1993b) *Investment Services in the Securities Field* ('Investment Services Directive'), Council Directive 93/22/EEC, May.

Fowle, M. (1991) '1992: Its Impact on European Banks, Their Structure, Operations and Accounts.' In R. Cranston (ed.) *The Single Market and the Law of Banking*, 2nd edition. Lloyds of London Press.

Group of Thirty (1993) *Derivatives: Practices and Principles*, July, New York.

Gruson, M. and Feuring, W. (1991) 'A European Community Banking Law: the Second Banking and Related Directives.' In R. Cranston (ed.) *The Single Market and the Law of Banking*, 2nd edition. London: Lloyd's of London Press.

Hall, M.J.B. (1989) 'The BIS Capital Adequacy "Rules": a Critique.' *Banca Nazionale del Lavoro Quarterly Review*, vol. 142: pp. 207–227.

Hall, M.J.B. (1991) 'The BCCI Affair.' *Banking World*, September: pp. 8–11.

Hall, M.J.B. (1995a) 'The Capital Adequacy Directive: an Assessment.' *Journal of International Banking Law*, vol. 10: pp. 78–87.

Hall, M.J.B. (1995b) 'The Measurement and Assessment of Market Risk: a Comparison of the European Commission and Basle Committee Approaches.' *Banca Nazional del Lavoro Quarterly Review*, vol. 194: pp. 293–330.

Roberts, R. (ed.) (1994) *International Financial Centres: Offshore Financial Centres*. Cheltenham: Edward Elgar.

Shavell, S. (1984a) 'Liability for Harm versus Regulation of Safety.' *Journal of Legal Studies*, vol. 13: pp. 357–374.

Shavell, S. (1984b) 'A Model of the Optimal Use of Liability and Safety Regulation.' *The Rand Journal of Economics*, vol. 15: pp. 271–280.

Steil, B. (1994a) 'International Securities Market Regulation.' In B. Steil (ed.) *International Financial Market Regulation*. Chichester: John Wiley.

Steil, B. (1994b) *International Financial Market Regulation*. Chichester: John Wiley.

14

The Proposed Additional Capital Requirements for the Management of Banks' Interest Rate Risk*

Paul Styger and Theo van Wyk

The 1970s was a boom period for banking in general, but also a time when financial markets became more competitive. This led to growing concern about risk management in the banking sector and about banking supervision in particular. The 1980s was, among other things, characterized by further dynamic changes to the financial system. The increase of international intermediation caused substantial balance sheet growth of banks in the industrialized countries. For example, the oil-producing countries began to increase their investment in the industrialized countries, whose banks then channelled the money to the developing countries. New, 'exotic' financial instruments were also introduced at an accelerated rate. One consequence of these developments was to increase the risks associated with banking activity (Llewellyn, 1988, p. 9). Although risks are inherent to banking, internationally there was growing concern about the increased level of these risks, especially the inadequate capital provisioning of banks. The purpose of bank capital is to absorb losses that could result from risky activities.

One of the most important risks faced by banks is interest rate risk, defined below. The interest rate determines a bank's cost of funds and the 'selling price' of its funds, in other words a bank's profitability. Interest rates also have an effect on a bank's liquidity, and the structure of its investment portfolio.

* Parts of this research were presented at seminars for bankers at the Had Hotel, Singapore, on 13 September 1994; The Bankers Club, Kuala Lumpur, Malaysia, on 14 September 1994; a session of the Biennial Conference of the Economics Society of Australia, Gold Coast, Australia, on 27 September 1994; and the NPI Conference, RAU, South Africa, on 27 October 1994. The authors wish to thank the participants for their helpful comments.

Recently the international banking community has begun to address the problems associated with the greater risks faced by banks, and the perceived inadequate level of capital the purpose of which is to absorb losses. This process began with the formation of committees by the Bank of International Settlements, and resulted in the Basle Concordat and later agreements inaugurated by the so-called Basle Committee. The work of these committees is an ongoing process, and since the end of 1993 one of the problem areas in banking that has been the subject of investigation is the nature of banks' exposure to interest rate risks.

Small states normally have relative small and open economies, and banks operating in these countries are, therefore, exposed to forces which to a large extent are outside their control, with the associated risks of unexpected interest rate changes. The proposed changes to the capital requirements for banks, which attempt to account for interest rate risk, are likely to have a major impact on the banks in these countries which are internationally active. They will also affect the role of banking supervision in these countries.

This chapter has several objectives. The first is to demonstrate some of the possible risks banks face in the management of expected interest rate changes. The second is to make some comments on the measures proposed by the Basle Committee for the management of exposure to interest rate risk. Finally, the chapter suggests an alternative method to the Basle Committee's proposal for identifying what the Committee terms 'outlier' banks; that is, banks which from the prudential safety perspective are perceived to have too large an exposure to interest rate risk.

We begin the chapter by presenting a short history of the Basle Committee. This is followed by a brief discussion of the management of interest rate risk by banks. We then discuss the results of some computer simulations that have been designed to demonstrate the risks involved in the management of interest rate changes. Finally we make some comments related to the Basle Committee's proposals for capital adequacy controls and interest rate risk management.

The Basle Committee

The Basle Committee was formed in 1974 in reaction to the problems (described above) that started in the banking industry in the 1970s (Steil, 1994; Bank Financial Management International, 1989: 7). The Committee was formed by the central banks and bank supervisors from the Group 10 countries, together with Switzerland and Luxembourg (Flint, 1992, p. 81).

The early work of the Basle Committee focused on the formulation of the agreement, known as the Concordat, that emphasized international cooperation between banking supervisors and the need to base bank supervision on a consolidated accounting basis (Compton, 1987, p. 92). In 1982 the Basle Committee became increasingly concerned over bank capital standards. In a report to the G10 countries it stated that the capital standards in international banking had been eroding and should be prevented from eroding further (Blanden, 1988, p. 56). In the following years it became clear that it was not enough simply to prevent the further erosion of bank capital; the structure of bank capital needed to be redefined in light of the Third World debt crises and other financial market innovations which had occurred in the international banking industry (Flint, 1992, p. 82).

In July 1988 the Basle Committee concluded an agreement establishing risk-based minimum capital standards for internationally active banks. The agreement became known as the Basle Accord and became fully effective at the end of 1992. The Basle Accord focuses primarily on credit risk; that is, the risk of a deterioration in the ability of a borrower or other counterparty to meet its obligations (Basle Committee, 1993, p. 4). It was adopted in all the major financial centres, both in and outside the G10 countries (Basle Committee, 1993, p. 4). It made provision for two forms of bank capital, tier 1 (primary) and tier 2 (secondary) capital, and a system of risk weights for the determination of the minimum capital standards. For purposes of calculating the minimum share capital and unimpaired reserve funds a bank is required to maintain, its assets and off-balance sheet activities are categorized according to their risk profiles and accorded applicable risk weights (on a standardized basis uniformly applicable to all banks) (Kelly, 1993, p. 274; Dewatripont and Tirole, 1994, p. 48f).

The Committee accepted that one major shortcoming of the Accord is that it does not impose minimum capital requirements for market-related banking risks such as interest rate risk. The Accord stated that the Committee would continue to study these banking risks and, where appropriate, incorporate them in a future market risk-based framework (Basle Committee, 1993, p. 4). This is discussed in the following section, with particular reference to South African banking practices.

Management of Interest Rate Risks

Interest rate risk is defined in this chapter as the risk that fluctuations in interest rates will unfavourably affect a financial institution's earnings and the value of its assets, liabilities and capital when such assets and liabilities mature, or are repriced at different times (Kelly, 1993, p. 320).

Banks in South Africa generally organize internal committees to manage the various risks to which they are exposed. To manage interest rate risk most of the major banks depend on asset and liability committees (ALCOs). The ALCO is responsible for shaping a bank's basic borrowing and lending strategy, by directing changes to the maturities and types of bank assets and liabilities in order to sustain profitability in a changing economic environment (Falkena *et al.*, 1989, p. 5). Asset and liability management focuses on controlling the net interest position of a bank, where the net interest is the

difference between the amount of interest received from loans and investments and the amount of interest paid on deposits and other liabilities (Kelly,
1993, p. 321).

In order to make a profit, banks normally run a mismatched book. Depending on interest rates and interest rate expectations, banks can borrow funds
over a relatively short period and grant loans for relatively long periods or
vice versa. If a bank has more (less) interest rate-sensitive assets than
liabilities maturing or being repriced in the near future (e.g. 90 days), it has a
positive (negative) gap (Falkena *et al.*, 1989, p. 48). The interest rate gap is,
therefore, a measure of the interest rate risk exposure of the bank. It also
serves as an important tool in managing interest rate risk.

Gap analysis implies that a bank's balance sheet is analysed. A specific
base date (commencement date of the gap period) is first selected. It is then
ascertained at what date after this base date every asset and liability on which
interest is received or paid will mature or become subject to reprising. A run-
off schedule can thereby be prepared. The ALCO can then prepare its
strategy to manage or control the interest rate risk of the bank's asset and
liability portfolio (Kelly, 1993, p. 330). The strategy involves two steps: first,
the prediction of future interest rate movements; second, the adjustments to
be made to the bank's portfolio in order to benefit from the expected changes
in the interest rate.

Clearly it is of crucial importance not only to predict interest rates very
accurately, but also to have some measure of the interest rate sensitivity of
the assets and liabilities. For the prediction of interest rate movements the
bank's perspective on term structure of interest rates is critical, and for the
measurement of interest rate sensitivity the concept of duration is important.
These two concepts will be discussed below, as they also feature in the
proposals arising from the Basle Committee.

Yield Curves and the Term Structure

The relationship between the yields on securities and their maturities across
the time spectrum is summarized in the term structure of interest rates. A
detailed discussion of the major theories that try to explain the term structure
of interest rates and their relationship to forecasting, such as the expectations
theory, the market segmentation theory and the liquidity preference theory
(see, for example, Cohen *et al.*, 1987; Horvitz and Ward, 1987), lies outside the
scope of this chapter.

In order to construct the yield curve, securities with similar features are
used for different periods of time. Moreover, only those instruments enjoying
high marketability are used to construct the yield curve (Van Gend, 1990, p.
71). It is worth noting that market expectations play a crucial role in the term
structure of interest rates. If interest rates are expected to rise, investors tend
to prefer short-term securities that represent an opportunity to benefit from a

future interest rate. Issuers of securities will prefer to borrow money long-term in order to fix the future funding costs at an expected lower level. The increase in the demand for short-term paper and the increased supply of long-term paper will cause the yield curve to become steeper or change shape. A positively shaped upward-sloping yield curve, where long-term interest rates are higher than the short-term rates, is the form of the yield curve that normally prevails. A negative yield curve, where short-term interest rates are higher than long-term interest rates, is considered to be an abnormal situation, and is expected to last for only a relatively short period of time.

Yield curves in South Africa over the past few years have changed from a positive to a negative, to a flat and back again to a positive. The latest negative yield curve in the case of South Africa lasted an unprecedented two years and nine months. From May 1989 to January 1992, short-term rates (e.g. 90-day liquid BA) were higher than long-term interest rates (e.g. ESKOM 168). As will be shown below, this situation can generate very large interest rate risks exposure for banks if it is not managed properly.

Duration and ALCO Activity

Duration is a measurement of interest rate risk exposure that has at its focal point price risk and reinvestment risk. Reinvestment risk occurs when cash flows from an instrument must be reinvested at a rate below that originally offered by the instrument. Price risk occurs when the market value of an instrument declines because of an increase in interest rates (Wyderko, 1988, p. 3).

Duration is the weighted average period within which all interest payments and principle of a fixed-interest security are received by its holder. The weights reflect the relative present value of the cash flows (Kelly, 1993, p. 99; Bierwag, 1987, pp. 57–60). To make the point more simply, the duration of a fixed-interest security is the time-weighted present value of the cash flows from the security divided by the present value of the cash flows. This concept is known as Macaulay duration. Modified duration is Macaulay duration divided by 1 plus the current interest rate (or yield).

As an interest rate risk management tool, duration attempts to measure the sensitivity of the market value of securities to changes in interest rates. The change in the market value of an investment (ΔMV) is given by (Wyderko, 1988, p. 4):

ΔMV = Duration x (change in interest rate) x (current market value) / (1 + current interest rate)

The duration of a portfolio of assets or liabilities can be obtained by calculating the duration of each security in the portfolio, and weighting the duration by the market value of the security as a percentage of the value of all assets or liabilities. The sum of the value-weighted durations is then the duration of the portfolio (Wyderko, 1988, p. 6). When the durations of the banks' assets and liabilities are unequal, the bank is exposed to interest rate risk. If the bank's asset duration exceeds its liability duration, it is said to possess a positive duration gap (Kelly, 1993, p. 335).

Simulation techniques are alternative methods which can be utilized to manage interest rate risk exposure. They are often used in conjunction with duration and gap analysis (Treasury Management International, 1994, p. 37). The results of simulation techniques refined for determining a certain South African bank's interest rate exposure will now be discussed.

Simulation Analysis for a South African Bank

One of the measures of the sensitivity of a bank's earnings (profit and loss) over the short term to a given change in interest rates which has been proposed by the Basle Committee is the so-called current earnings perspective (Basle Committee, 1993, p. 10). According to the Basle Committee (1993, p. 10), the main focus of the current earnings perspective is the sensitivity of a bank's profit and loss account in the short term (one year) to a given change in interest rates. This sensitivity is measured by the duration of the deposits and assets. The objective of the Basle Committee's proposals is to identify banks that are especially vulnerable to fluctuations in recorded profits. This was the approach that we adopted in the following simulations.

For the purpose of the simulations, we utilized the data of a South African bank, which in some countries would be described as a savings and loan institution.

The simulations were undertaken using the ALMAN system, Version 4.57 of SPL (Systems Programming Ltd, 1994a), and the interest rate forecasts and yield curves for the simulations were obtained from the IFS (Interest Rate Forecasting System), Version 1.83 (Systems Programming Ltd, 1994b). This system is being used by the treasuries of the majority of South African banks, as well as by a number of countries outside South Africa. (More specific details of the simulation can be provided by the authors on request.)

Simulation 1

For the first simulation it was assumed that the bank forecast that a normal (positive) yield curve would prevail for the 12 months of the forecasting horizon, and structured its borrowing and lending strategy accordingly in line with previously outlined criteria. In simulation 1, the assumption is made that the bank's forecast about the shape of the yield curve turned out to be accurate. Positive yield curves were used for this simulation.

Simulation 2

In the second simulation it was again assumed that the bank forecast a positive yield curve, with the same borrowing and lending strategy as in simulation 1. However, in this simulation its forecast was not accurate, and negative yield curves were used in this simulation to depict the 'true' interest rate term structure.

Simulation 3

In the third simulation it was again assumed that the bank forecast a positive yield curve, again adapting the same borrowing and lending strategy as in simulation 1. Once again, in this simulation its forecast was inaccurate, and flat yield curves were used in this simulation to depict the 'time' interest rate term structure.

Simulation 4

In the fourth simulation the same interest rate assumptions as in simulation 1 were made and those forecasts are again assumed to be accurate (*ex post*). The difference between simulation 4 and simulation 1 is that in simulation 4 the bank continuously reassessed its borrowing and lending strategy to take advantage of the extent to which the yield curve turned positive over the forecast horizon.

As discussed previously, the ALCO of the bank is assumed to focus on the net interest income. The ALMAN system prepares detailed income statements of net interest income. In the discussions below, the detailed income statements will not be shown, but only the net interest income of every simulation (see Tables 14.1 and 14.2 – detailed income statements can be obtained from the authors on request.)

Because the change in the shape of the yield curve takes place over time, there is initially some degree of similarity in the results, as is indicated by the short rates for the positive and negative yield curves over the three- and six-month forecast periods. That is also the main reason why, when the yield curve turned out to be negative instead of positive (simulation 2), it did not affect the net interest income of the bank too adversely (see Table 14.1). The flat yield curve proved to be more problematic. Although the flat yield curve does not represent a change in the relationship between short- and long-term interest rates, it does imply a movement in interest rates away from those expected, and thus has an effect analogous to that of an interest rate change. Furthermore, the flat yield curve need not remain at exactly the same interest rate levels, meaning that the yield curve can move either up or down – a parallel-type shift – with either long- or short-term interest rates moving marginally more than the other. Overall, however, the yield curve stays flat.

Table 14.1   Simulations 1 to 3: income statement

	Finance charges paid			Interest received			Net interest income		
	Flat	*Negative*	*Normal*	*Flat*	*Negative*	*Normal*	*Flat*	*Negative*	*Normal*
P 1	3,562,470	4,081,867	4,062,584	5,191,985	5,243,593	5,242,191	1,629,516	1,161,725	1,179,608
P 2	3,485,332	3,997,180	3,974,851	4,493,102	5,089,908	5,089,190	1,007,770	1,092,728	1,114,338
P 3	3,616,058	4,151,373	4,126,431	4,659,354	5,283,653	5,282,948	1,043,296	1,132,280	1,156,517
P 4	3,549,315	4,266,863	4,239,219	4,726,672	5,311,158	5,309,828	1,177,358	1,044,296	1,070,609
P 5	3,351,116	4,230,999	4,185,147	4,518,600	5,299,424	5,298,052	1,167,484	1,068,425	1,112,905
P 6	3,372,594	4,482,163	4,419,568	4,555,649	5,631,990	5,630,564	1,183,055	1,149,827	1,210,996
P 7	3,237,650	4,398,659	4,330,336	4,322,147	5,610,846	5,609,785	1,084,496	1,212,187	1,279,448
P 8	3,304,431	4,602,142	4,524,573	4,371,598	5,989,179	5,988,441	1,067,167	1,387,036	1,463,869
P 9	3,276,925	4,657,612	4,568,791	4,405,908	6,038,382	6,037,567	1,128,983	1,380,770	1,468,777
P10	2,923,856	4,258,461	4,170,930	3,883,251	5,494,396	5,493,432	959,395	1,235,936	1,322,502
P11	3,185,670	4,754,308	4,640,120	4,300,491	6,217,099	6,216,016	1,114,821	1,462,791	1,575,896
P12	3,049,329	4,661,491	4,527,761	4,057,446	6,071,347	6,070,214	1,008,117	1,409,857	1,542,453

Note: P = period.

Table 14.2   Simulation 4: income statement managed strategy (rate view: normal)

	Finance charges paid	Interest received	Net interest income
P1	4,032,994	5,242,191	1,209,198
P2	3,956,673	5,089,930	1,133,257
P3	4,108,267	5,284,405	1,176,138
P4	4,212,425	5,311,332	1,098,906
P5	4,150,535	5,298,723	1,148,188
P6	4,372,077	5,630,387	1,258,310
P7	4,286,352	5,610,039	1,323,688
P8	4,482,546	5,988,531	1,505,985
P9	4,525,428	6,037,321	1,511,893
P10	4,142,550	5,494,153	1,351,604
P11	4,608,823	6,216,826	1,608,003
P12	4,486,804	6,069,967	1,583,163

Note: P = period.

Table 14.3   Durations of simulations 1 to 4

Simulations	Period 1		Period 12	
	Assets	*Deposits*	*Assets*	*Deposits*
1	5.01	1.40	4.66	0.97
2	5.01	1.40	4.66	0.96
3	5.57	1.56	6.61	1.57
4	5.01	1.41	4.56	0.95

That is what happened in simulation 3 (Table 14.1). Simulation 4 demonstrates, as expected, that the management of the interest rate exposure on a month-to-month basis achieved better results (Table 14.2).

The ALMAN system computes a complete duration report for all deposits and assets. In the results shown in Table 14.3, only the weighted totals of the duration for the assets and deposits are shown.

In terms of net interest income, the duration analysis confirms the results reported above. It shows that the deposits and assets under simulation 3 were more sensitive to a change in the interest rate than those of any of the other simulations, as time progressed.

Mere duration does not, however, seem to be enough to identify what the Basle Committee (1993, p. 10) calls an 'outlier' bank. The banks submit only the duration value of the beginning of the evaluation period. This in itself has little meaning in this context, owing to the fact that static results do not give enough information for the banking supervisors to identify so-called 'outlier' banks. It is only with a simulation where the change in duration over time is analysed (see Table 14.3) that more conclusive figures become available. This in itself does not resolve the problem of standardizing each bank's simulation results, as each bank has its own portfolio management strategy.

Duration-Weighted Assets

An alternative method that could yield a better indication of 'outlier' banks to the banking supervisors is to use duration-weighted asset (DWA) values. The crux of the Basle Accord was the imposition of capital requirements as assets. In the same manner of thinking, 'outlier' banks, specified in terms of their interest rate risk, can be identified by the percentage change in their DWAs for a specific interest rate change (e.g. of 1 per cent) over a specific period of time. For the purposes of calculation, the following assumptions are made (these should in practice be supplied to the banks by the banking supervisors):

- The calculations are undertaken for a 1 per cent change in interest rates.
- The base rate for the calculations is the average short-term yield rate.

The DWAs for the specific period were calculated as follows:

$$DWA = (-D(\Delta r)(A))/(1+r)$$

where D is the duration of the assets for the specific period, Δr is the change in the interest rate (1 per cent), $(1+r)$ is (1+ the base interest rate) – in this example $r = 16$ per annum, and A is the rand value of the assets for the specific period.

This approach can be of assistance to the banking supervisors in identifying 'outlier' banks, because it gives a better indication of the banks' interest

Table 14.4   The duration-weighted assets (DWAs) for simulations 3 and 4

Simulation	DWA period 1	DWA period 2	% change
4	1,426,732,000	1,405,183,000	-1.51
3	1,590,086,000	1,977,376,000	24.36

rate exposure. As indicated earlier, this does not, however, solve the problem of standardizing banks' simulation results. If the banks were to report the percentage change in their DWAs (i.e. the change in the market value) which would arise for a 1 per cent change in the interest rate (even though the interest rate did not change) from one report period to the next, the supervisors would be able to use this to identify 'outlier' banks. The advantage of this approach is that they do not work on the basis of simulated results, but still use real-time, dynamic comparisons. The disadvantage is that such an approach utilizes *ex post* results, and on the basis of these, the banks and their supervisors cannot take precautionary steps to reduce the interest rate exposure (*ex ante*). So by the time action is taken it may be too late.

Conclusion

The purpose of this chapter was to demonstrate some of the possible risks arising for a bank in the management of expected interest rate changes, and to make some comments on the measures proposed by the Basle Committee to manage the bank's exposure to interest rate risk. We also attempt to propose an alternative method to that of the Basle Committee to identify 'outlier' banks.

It is shown that interest rate forecasting is critical to banks' risk management, thereby demonstrating that the Basle Committee's concern over interest rate exposure is well founded. However, we also demonstrate that the proposal to use duration as a measure to identify 'outlier' banks has a number of (serious) limitations. This chapter shows that some of these limitations may be overcome by the use of appropriately designed simulations and the concept of duration-weighted assets.

Our conclusions have wide-ranging implications for the financial section in small states. As stated at the start of the chapter, these countries tend to face unexpected pressures on interest rates, owing to the very high degree of economic openness, and thus wild fluctuations in interest rates, causing rapid adjustments in both the shape and position of the yield curve. The proposed changes to the capital requirements for banks, which attempt to account for interest rate risk, are therefore likely to have a major impact on the banks in those countries which are internationally active. They will also affect the role of banking supervision in these countries.

References

Bank Financial Management International (1989) 'Basle Supervisor Defends Secret Reserves,' November: pp. 6–9.

Basle Committee on Banking Supervision (1993) *Measurement of Banks' Exposure to Interest Rate Risk: Consultative Proposal by the Basle Committee on Banking Supervision*. Basle: Bank for International Settlements.

Bierwag, G.O. (1987) *Duration Analysis: Managing Interest Rate Risk*. Cambridge, MA: Ballinger.

Blanden, M. (1988) 'Ironing Out Those Troublesome Bumps.' *The Banker*, vol. 138: pp. 56–59.

Cohen, J.B., Zinbarg, E.D. and Zeikel, A. (1987) *Investment Analysis and Portfolio Management*, 5th edition. Homewood, IL: Irwin.

Compton, E.N. (1987) *The New World of Commercial Banking*. Lexington, MA: Lexington Books.

Dewatripont, M. and Tirole, J. (1994) *The Prudential Regulation of Banks*. Cambridge, MA: MIT Press.

Falkena, H.B., Kok, W.J. and Meijer, J.H. (1989) *Financial Risk Management in South Africa*. Cape Town: Macmillan.

Flint, J.H.L. (1992) Bankkapitaalvereistes na aanleiding van die Basel-ooreenkoms. Unpublished MComm thesis, Potchefstroom University for CHE.

Horvitz, P.M. and Ward, R.A. (1987) *Monetary Policy and the Financial System*, 6th edition. Englewood Cliffs, NJ: Prentice-Hall.

Kelly, M.V. (1993) *Financial Institutions in South Africa: Financial, Investment and Risk Management*. Cape Town: Juta.

Llewellyn, D.T. (1988) 'The Policy Framework: Capital Adequacy.' In Wilson, J.S.G. (ed.) *Managing Bank Assets and Liabilities*. London: Euromoney Publications, pp. 7–54.

Steil, B. (1994) *International Financial Market Regulation*. Chichester: J. Wiley & Sons.

Systems Programming Ltd (1994a) *ALMAN: SPL's Asset/Liability Management System*, Version 4.57. Rivonia: SPL.

Systems Programming Ltd (1994b) *IFS: An Interest Rate Forecasting System*, Version 1.83. Rivonia: SPL.

Treasury Management International (1994) 'Interest Rate Products,' vol. 19: pp. 35–40.

Van Gend, G.O. (1990) 'The Money Market Yield Curve.' Unpublished BComm. Hons thesis, University of Port Elizabeth.

Wyderko, L.W. (1988) *A Duration Analysis Primer*. Rolling Meadows, IL: Bank Administration Institute.

15

The 'OECD Club Rule' in Bank Risk-based Capital Adequacy: Implications for Malta and Other Small States

Marcel Cassar

In 1988 the Basle Committee on Banking Supervision published its report *International Convergence of Capital Measurement and Capital Standards* – known as the 'capital accord' – which established a basic regulatory framework for bank risk-based capital levels. While it has been generally held that the Basle rules have caused bankers to appreciate, and focus on, the need to measure and provide for the risk involved in their lending and investing activities, the detail of the framework has met with mixed reactions and strong views that the risk-based formula has significant shortcomings which undermine its usefulness.

This chapter explores one particularly controversial area of the accord which the Basle Committee is under pressure to re-examine: the so-called 'OECD[1] Club rule'. It will evaluate the claims by bankers that the rule is to blame for certain market distortions and assess whether, as a result, it effectively hinders the capital accord from achieving its level-playing-field objectives. The rule's implications for the financial operations of entities operating in small states will then be examined, with particular reference to the case of Malta. This is followed by suggested alternative approaches to the rule which would still reflect country risk-based principles.

Certain Shortcomings of the Risk Weightings

The 1988 accord appeals because of the simplicity of the framework adopted, in which only given weights are used: 0, 10, 20, 50 and 100 per cent. However, the Basle Committee warns that 'there are inevitably some broad-brush judgements in deciding which weight should apply to different types of asset and the weightings should not be regarded as a substitute for commercial judgement for purposes of market pricing of the different instruments' (Basle Committee, 1988, para. 29). Hall (1989) questions this observation and remarks that 'given the capital constraints likely to be faced by banks as a result of the impositions of its [Basle's] proposals, it is not clear that banks have much room for manoeuvre vis-à-vis pricing in a highly-competitive world'.

The derivation of the weightings themselves has been subject to strong criticism. Basle's main preoccupation has been with credit risk, although interest rate and market risks have now been treated in the more recent consultative proposals issued by the Committee in 1994. It is also arguable how standard weightings can adequately reflect credit risk without this taking into account the particular and specific characteristics of the counter-party. Hall's point that 'it is not clear that the relativities between different assets components have been correctly established' is best exposed by questioning how realistic it is that a private-sector claim is five times as risky as a domestic inter-bank loan.

More fundamentally, the inability of the risk weightings to reflect the true underlying risk potential of the assets to which they are assigned is a major shortcoming to which criticism is frequently addressed. Two banks with similar risk–asset compositions (and therefore allocating the same amount of capital to these assets) may have widely different asset quality profiles. For a risk-based ratio to become a truly risk-adjusted ratio requires consistency in asset quality monitoring, otherwise the true capital strengths of banking institutions would be misrepresented.

Lomax (1987) has pointed out that

> the assignation of risk weights to different balance sheet activities (and conversion factors to off-balance-sheet activities) fundamentally affects a bank's business strategy, pricing policy and capital allocation; failure to reflect 'true' risk in risk weights therefore leads to important distortions in business policy and resource allocation, with concomitant effects for consumers and the real economy.

Perhaps this is where the risk weightings present their greatest short-comings, and critics of the OECD Club rule have been claiming that such undesirable effects actually result from the application of the formula.

Origins of the OECD Club Rule

The OECD Club rule is the term applied to the methodology introduced into the capital accord for determining country transfer risk. In 1988, memories of the less developed countries debt crisis were very vivid in the minds of international banking regulators, and the Committee wanted to make a distinction between debt-defaulting countries and other countries.

In its earlier consultations with banking supervisors, the Basle Committee had suggested two alternative approaches for incorporating country transfer risk into the framework. One consisted of 'a simple differentiation between claims on domestic institutions (central government, official sector and banks) and claims on all foreign countries' (Basle Committee, 1988). The alternative was to differentiate 'on the basis of an approach involving the selection of a defined grouping of countries considered to be of high credit standing' (ibid.). During the consultative period, banks and banking associations in the G10 countries showed themselves as overwhelmingly in favour of the second alternative. Their most important argument against the first alternative was that the member states of the European Community (comprising seven of the G10 countries) were committed to the principle that all claims on banks, central governments and the official sector within the Community should be treated in the same way, and there would be difficulties and an 'undesirable asymmetry' in the manner in which the seven EC countries would apply the domestic–foreign split compared with the non-EC G10 countries.

The Basle Committee therefore concluded that

> a defined group of countries should be adopted as the basis for applying differential weighting coefficients, and this group should be full members of the OECD or countries which have concluded special lending arrangements with the IMF associated with the Fund's General Arrangements to Borrow. (para. 35)

The consequences of the OECD Club rule for the weighting profile can be summarized as follows:

- Claims on OECD central government or public-sector entities (PSE) attract a zero or low weighting: claims on non-OECD central government or PSEs attract such weighting only if denominated in local currency and funded in liabilities in that same currency, otherwise a high weighting is applied.
- Short-term inter-bank claims with the residual maturity of up to and including one year carry a 20 per cent weighting, irrespective of country of incorporation; longer-term claims on OECD-incorporated banks are weighted at 20 per cent while longer-term claims on non-OECD-incorporated banks are weighted at 100 per cent.
- The risk methodology outlined in the two points above is also reflected with regard to claims carrying the guarantees of, or secured by collateral in the form of instruments issued by, OECD central governments, PSEs and incorporated banks.

Criticisms of the Rule

The rule effectively requires a bank to maintain a higher capital requirement when it is exposed to a country not included in the OECD grouping, with the result that such lending becomes more costly. Critics of the rule range from those expressing concern that non-OECD banks are adversely affected when tapping the market for funds, to more controversial statements that such 'practices and regulations ... bias banks in industrial countries against lending to developing countries by requiring additional conditions that restrict voluntary credit and investment flows to developing countries' (G24 Developing Countries meeting, 1991; see Deutsche Bundesbank, 1991). Truly, there has been a certain concern that as banks take their credit decisions largely on the basis of the impact on the risk–asset ratio, the OECD Club rule significantly affects the pricing of such decisions since it affects the banks' cost of capital.

The criticisms levelled at the rule are many and wide-ranging. First, the question arises as to what validity the rule can hold if it is purportedly founded on a desire to distinguish between creditworthy countries and debt defaulters. The anti-rule lobby includes two supervisory groups affiliated to the Basle Committee, namely the Arab Committee on Banking Supervision and the Offshore Group of Banking Supervisors. Members of these two interest groups, which include the Gulf states and territories such as Hong Kong, Bermuda and the Channel Islands respectively, have criticized the rule as being discriminatory against those non-OECD countries with a record of not having any debt, and therefore incapable of being identified as debt defaulters. On the other hand, certain OECD countries carry a high debt burden.

This leads to a second criticism: the question as to whether by creating an OECD–non-OECD divide the Basle Committee was actually seeking to base its distinction on the quality of prudential regulations and supervisory practices in place in particular jurisdictions. Greater cooperation between regulators in recent years has led to improved harmonization and tighter supervisory practices across borders and, as a result, an increased number of non-OECD countries have established high standards of regulation and supervision which are certainly not inferior to those of most OECD member countries. One can mention the example of Singapore, a country outside the OECD which has a reputation for strict regulation and which has adopted into its legislation a minimum risk–asset ratio of 12 per cent based solely on core capital – far more stringent than the situation obtaining in most OECD countries.

Third is the related issue of how membership of the OECD is recognized for the purposes of the Club rule; that is, whether territories related to members of the OECD which consider themselves as bound by the OECD convention (e.g. the Channel Islands and the Isle of Man/United Kingdom) could be regarded as being 'inside the club'. It is possible that there is no coherent approach among Committee members with regard to this problem, with the result that the OECD Club rule further deviates from the objective of achieving a level playing-field.

Fourth, the rule is often criticized as being simple and convenient, but only on the basis of arbitrary solutions. This goes directly against the principle underlying the capital accord of identifying and assessing specific risk situations. It is difficult to envisage long-term support for a rule suspected of being founded on arbitrary arguments, possibly reflecting political pressures from the positions taken by the industrialized nations.

The widening of the OECD group definition to include 'countries which have concluded special lending arrangements with the IMF' meant that Saudi Arabia, from which the Fund borrowed short-term in 1981, became the only non-OECD country to qualify for a zero- or low-risk weighting. However, that was not well received by the other oil-rich countries.

Furthermore, it is possible that the rule may carry implications for the strategic direction and positioning of non-OECD banks. For example, the incorporation as a UK company of HSBC Holdings plc – the parent of Hongkong and Shanghai Banking Corporation – can be as much attributed to the future of Hong Kong after 1997 as to the business implications of remaining incorporated outside the OECD.

The rule, however, has also had its share of defence. Huib Muller, then Chairman of the Basle Committee, when referring to the weighting for claims on non-industrialized countries, remarked:

> This was not, as is sometimes claimed, discrimination against such claims; many banks would regard a short-term claim on, say, Royal Dutch Shell as inherently less risky than a 20-year loan to the government of a developing country, yet they are weighted the same ... ' (Muller, 1989, p. 5)

Proponents of the rule also observe that while Basle Committee pronouncements are statements of best supervisory practice, European Union directives are legally binding on all member states. The decision by the European Commission to adopt an approach similar to the OECD Club rule when promulgating the 'Solvency Ratio Directive' (EC, 1989) renders the likelihood of fundamental modifications to the rule more remote. It is also to be acknowledged that an alternative to the Club rule for reasonably measuring country transfer risk has been elusive, although pressures have been mounting for the rule to be reviewed and replaced by a more suitable weighting basis.

Implications for the Borrowing and Lending Activities of Operators in Small States – the Case of Malta

It is relevant to point out that, as a direct result of the OECD Club categorization, claims on a large number of small states with excellent standards of regulation, outstanding economic performance and, possibly, an absence of debt record are assigned a higher risk weighting. The reference above to offshore financial centres suggests that operators incorporated in a number of non-OECD states with a successfully developed and thriving offshore financial services industry, and typically engaged in international financial operations, are particularly interested in the implications of the rule.

Malta is not a member of the OECD, hence international banks would be required to apply a high risk weighting to Maltese exposures. However, any suggestion that the borrowing experiences of Maltese entities have been adversely affected by the operation of the Club rule is far from supported by evidence. On the contrary, the recent borrowings by Maltese companies in the international financial markets invariably followed a generally positive credit evaluation. In the case of Enemalta, the Maltese state-owned energy corporation, which raised a US$100 million government-guaranteed syndicated loan late in 1992, the *Financial Times*, while observing that 'Maltese borrowings are 100% weighted', described the basic yield as 'not over-generous' (Bollen, 1992). Bank of Valletta Ltd, in mandating a US$40 million syndicated credit facility in 1993 without recourse to a government guarantee, received a favourable evaluation of its credit standing by the participating banks. Also, the 1994 government-guaranteed global Eurobond issue of US$205 million by the Freeport Terminal (Malta) Ltd was preceded by the assignation of Malta's first-ever credit rating of an 'A-2' from Moody's and an 'A' from Standard and Poor's, both favourable ratings for Malta. Finally, the recent Ecu 60 million borrowing by Air Malta for the purchase of three RJ70 Avroliner aircraft was supported by a guarantee from the UK export credit agency ECGD.

There are of course other, possibly overriding, factors that have an important influence on the outcome of the determination of the final terms and conditions in any borrowing arrangement. While it would be unfounded to suggest – indeed, such an exercise would be outside the scope of this chapter – that Maltese banks and corporations have been disadvantaged in their international borrowing operations as a direct result of the application of the Club rule, it must also be acknowledged that the burden of a high risk weighting is a factor which borrowing entities will undoubtedly want to neutralize.

There are also implications with regard to the overseas lending activities of Maltese incorporated banks. Their proximity to the emerging markets of North Africa and the Near East heightens their potential to develop project and corporate finance business. However, the banks' competitive positioning could be adversely affected if their cost of funds – a vital element in the case of financing activities typically backed by wholesale funding – were to increase as a result of the workings of the rule. Moreover, their cost of capital would be raised since exposures to these markets would typically be 100 per cent weighted. The creation of a development bank for the region, as well as the arrival of banks to which the risk-based capital adequacy standards do not apply, may lead to a further 'unlevelling' of the field of operations.

The absence of a firm stand against the OECD Club rule may further weaken the position which Malta should be taking in the interest of its aspirations to develop as an international financial centre for investment business. For example, the Japanese Securities Dealers' Association requires mutual funds to be established in an OECD member country if that fund is to be marketed to the Japanese public. Other countries operate similar criteria.

Alternatives to the OECD Club Rule

Any suggestion that the Club rule be abandoned begs the question of what suitable alternative methodology can be devised to take into account country transfer risk. One alternative is to allow home-country bank supervisors to determine country risk by reference to principles which the Basle Committee would delineate. Such an approach could be assisted by the introduction of a wider band of risk weights – the number could be increased from, say, five to ten – which counterparties (borrowers) would be induced to progressively improve.

Another alternative would be to have a club definition based on a more comprehensive approach to determining the creditworthiness of a country and its official and financial institutions. This could be derived from such official development criteria as are established by the IMF and World Bank; and less official, but equally useful, releases by the financial press such as *Institutional Investor's* authoritative semi-annual credit survey. For example, in the March 1994 country credit rating rankings, Singapore and Taiwan, both non-OECD members, were placed 11th and 12th respectively, ahead of a number of OECD members including Iceland and Turkey, which were placed 32nd and 45th respectively. A more formal and supranational country risk rating framework could be adopted on an initiative that could be assumed by an international organization such as the Bank for International Settlements.

But 'bank supervisors are not in the business of promoting aid for struggling countries . . . nor should they set themselves up as rating agencies' (*The Banker*, December 1991, p. 7). Hence the other suggestion that each individual country should be left to define its own 'club' for its banks, taking into consideration specific political realities, e.g. whether Hong Kong and Malaysian banks could have Chinese exposures as 100 per cent weighted. Supervisors could also establish country limits for individual banks which are linked to their capital base. In this way, countries perceived as better risks can be allocated limits equivalent to increasingly higher percentages of a bank's capital.

A modification would be to adopt a checklist approach on the lines of the Bank of England's matrix for provisioning against sovereign debt. A similar approach could be developed from the idea of the Comptroller of the Currency's risk-based capital matrix: a scoring system would guide bankers through a series of credit quality indicators in order to arrive at a final total score which would then be converted into a risk weighting. One should, however, bear in mind the overriding consideration of simplicity in adopting such approaches.

One particular concern with the OECD Club rule is the suggestion that any new OECD members are automatically allowed to join the Club, irrespective of their risk status. This has recently prompted the Basle Committee to effect a small but significant revision, effective January 1995, of its definition of the OECD Group in the 1988 accord. To paragraph 35 has now been added: 'Any country which reschedules its external sovereign debt is, however, precluded from the defined group for a period of five years' (Basle Committee, 1994). On this development *Thomson's* (1994, p. 3) has commented:

> although an official source said the change is not aimed at any country, Mexico's entry into the OECD earlier this year appears to be the principal cause of the change. Mexico, which is widely considered to have triggered the debt crisis in 1982, is the only country in the group that has rescheduled its debt within the last five years.

Moreover, it is also likely that the Committee was aiming to anticipate the possibility of other admissions into the OECD in the near future.

Conclusion

Whichever methodology is adopted – and there is no 'best' system – the underlying objective should remain that of reflecting a country's ability to service its external obligations and making available the conditions to allow private-sector operators to do likewise. This evidently suggests that economic and political as well as qualitative factors need to be taken into account.

Peter Cooke, Chairman of the Basle Committee at the time the capital accord was promulgated, describes the Club rule as

> a very broadly based and somewhat arbitrary distinction between countries to incorporate within the framework some differentiation between different degrees of country risk. This distinction was in view of all the banking associations in G10 countries a necessary feature if the framework as a whole were to have credibility. (Cooke, 1990)

The key words in this statement are *differentiation* and *credibility*, two necessary elements which must go together if the risk-based capital adequacy regime is to survive with authoritativeness.

Although the recent revision by Basle of the OECD definition should go some way to allaying the concerns of the critics, it still leaves unresolved the Club rule's main problem of being discriminatory against non-OECD countries. As claims to the effect that the Club rule causes market pricing distortions become more resounding and possibly substantiated by evidence, anti-rule lobbyists might be in a stronger position than ever should they register their complaint by invoking the inconclusive and yet-to-reconvene WTO agreement on financial services.

In the meantime, the OECD Club rule will continue to draw more attention as analysts harness their efforts in an attempt to establish whether the risk weightings in a capital formula designed to promote harmonization of capital adequacy regulation are frustrating the achievement of the fundamental goal of a supervisory level playing-field. But more important will be the search for evidence that the rule is having a significant effect on the cost of developing nations' borrowing.

Note

1. The Organisation for Economic Co-operation and Development (OECD) was created in 1961. Its present members are Australia, Austria, Belgium, Canada, Denmark, Finland, France, Germany, Greece, Iceland, Ireland, Italy, Japan, Luxembourg, Mexico, the Netherlands, New Zealand, Norway, Portugal, Spain, Sweden, Switzerland, Turkey, the United Kingdom and the United States.

References

Basle Committee on Banking Regulations and Supervisory Practices (1988) *International Convergence of Capital Measurement and Capital Standards*. Basle: Bank for International Settlements.

Basle Committee on Banking Supervision (1994) *Amendment to the Capital Accord of July, 1988*. Basle: Bank for International Settlements.

Bollen, B. (1992) '$100m Maltese debut in syndicated loan markets.' *Financial Times*, 26 November.

Cooke, P. (1990) 'International Convergence of Capital Adequacy Measurement and Standards.' In E.P.M. Gardener (ed.) *The Future of Financial Systems and Services*. London: Macmillan.

Deutsche Bundesbank, Auszüge aus Presseartikeln (1991) *Communiqué of the Intergovernmental Group of Twenty-four on International Monetary Affairs*. 45th meeting in Bangkok, Thailand.

EC (1989) *Solvency Ratio for Credit Institutions* ('Solvency Ratio Directive'), Council Directive 89/647/EEC, December.

Hall, M.J.B. (1989) 'The BIS Capital Adequacy "Rules": a Critique.' *Banca Nazionale del Lavoro Quarterly Review*, June: pp. 218–219.

Lomax, D. (1987) 'Risk Asset Ratios: a New Departure in Supervisory Policy.' *National Westminster Bank Quarterly Review*, August.

Muller, H.J. (1989) 'The Capital Agreement of the Basle Committee in a European Context.' Journée d'Études Financières symposium on 'The Management of Banking Risk and Resources,' Luxembourg, 23 November.

Thomson's International Banking Regulator (1994) 'Basle Committee Revises OECD Definition.' American Banker Inc., 8 August.

Author Index

Subject Index